EURASIA

The global political map is undergoing a process of rapid change as former states disintegrate and new states emerge. Territorial change in the form of conflict over land and maritime boundaries is inevitable but the negotiation and management of these changes threaten world peace.

Eurasia offers a wide-ranging and original interpretation of territory, boundaries and borderlands in Europe, Asia and the Far East.

World Boundaries is a unique series embracing the theory and practice of boundary delimitation and management, boundary disputes and conflict resolution, and territorial change in the new world order. Each of the five volumes – *Global Boundaries, The Middle East and North Africa, Eurasia, The Americas* and *Maritime Boundaries* – is clearly illustrated with maps and diagrams and contains regional case studies to support thematic chapters. This series will lead to a better understanding of the means available for the patient negotiation and peaceful management of international boundaries.

Carl Grundy-Warr is Lecturer in Geography at the National University of Singapore and Research Associate at the International Boundaries Research Unit, University of Durham. He is also a member of the Association of Borderlands Scholars, USA.

WORLD BOUNDARIES SERIES
Edited by Gerald H. Blake
Director of the International Boundaries Research Unit
at the University of Durham

The titles in the series are:

GLOBAL BOUNDARIES
Edited by Clive H. Schofield

THE MIDDLE EAST and NORTH AFRICA
Edited by Clive H. Schofield and Richard Schofield

EURASIA
Edited by Carl Grundy-Warr

THE AMERICAS
Edited by Pascal Girot

MARITIME BOUNDARIES
Edited by Gerald H. Blake

EURASIA

World Boundaries volume 3

Edited by Carl Grundy-Warr

London and New York

First published 1994
by Routledge
2 Park Square, Milton Park, Abingdon, Oxon, OX14 4RN

Simultaneously published in the USA and Canada
by Routledge
270 Madison Ave, New York NY 10016

Reprinted 1997

Transferred to Digital Printing 2007

Typeset in 10pt September by Solidus (Bristol) Limited

British Library Cataloguing in Publication Data
A catalogue record for this book is available from the British Library

Library of Congress Cataloguing in Publication Data
A catalogue record for this book
is available from the Library of Congress.

ISBN 0-415-08834-8
5-vol. set: ISBN 0-415-08840-2

Publisher's Note
The publisher has gone to great lengths to ensure the quality
of this reprint but points out that some imperfections in the
original may be apparent

CONTENTS

CONTENTS

Part II Asia-Pacific

FIGURES

TABLES

NOTES ON CONTRIBUTORS

Francis M. Auburn is Deputy Director of the Centre for Commercial and Resources Law, University of Western Australia, Nedlands, Australia.

Jeremy Black is Reader in History at the University of Durham, United Kingdom.

Hanns Buchholz is Director of the Institute of Regional Geography at Leipzig and he teaches at the Department of Geography, Hanover University, Germany.

Vivian Forbes is Marine Geographer at the University of Western Australia, Nedlands, Australia.

Michel Foucher is Director of L'Observatoire Européen de Géopolitique, Lyon, and he teaches at the Université de Lyons II (Lumières), France.

Patricia Goodstadt is a doctoral researcher at the School of Geography, University of Oxford, United Kingdom.

Carl Grundy-Warr teaches at the Department of Geography at the National University of Singapore.

Kimie Hara is a researcher on the Program on International Economics and Politics, East–West Center, Honolulu, Hawaii, United States of America.

Julian V. Minghi teaches at the Department of Geography, University of South Carolina, Columbia, United States of America.

Nurul Islam Nazem teaches at the Department of Geography, Jahangirnagar University, Dhaka, Bangladesh.

Dennis Rumley is an Associate Professor at the Department of Geography, University of Western Australia, Nedlands, Australia.

John Scott is a geological consultant for the petroleum industry, and was formerly Director of the Western Australia Center for Petroleum Exploration, Curtin University of Technology, Bentley, Australia.

Peter Kien-hong Yu teaches at the Sun Yat-sen Institute, National Sun Yat-sen University, Kaohsiung, Taiwan and is a Research Fellow at both Asia and World Institute and Chinese Council of Advanced Policy Studies in Taipei.

SERIES PREFACE

The International Boundaries Research Unit (IBRU) was founded at the University of Durham in January 1989, initially funded by the generosity of Archive Research Ltd of Farnham Common. The aims of the unit are the collection, analysis and documentation of information on international land and maritime boundaries to enhance the means available for the peaceful resolution of conflict and international transboundary co-operation. IBRU is currently creating a database on international boundaries with a major grant from the Leverhulme Trust. The unit publishes a quarterly *Boundary and Security Bulletin* and a series of *Boundary and Territory Briefings*.

IBRU's first international conference was held in Durham on 14–17 September 1989 under the title of 'International Boundaries and Boundary Conflict Resolution'. The 1989 conference proceedings were published by IBRU in 1990 under the title *International Boundaries and Boundary Conflict Resolution*, edited by C.E.R. Grundy-Warr. The theme chosen for our second conference was 'International boundaries: fresh perspectives'. The aim was to gather together international boundary specialists from a variety of disciplines and backgrounds to examine the rapidly changing political map of the world, new technical and methodological approaches to boundary delimitations, and fresh legal perspectives. Over 130 people attended the conference from 30 states. The papers presented comprise four of the five volumes in this current series (Volumes 1–3 and Volume 5). Volume 4 largely comprises proceedings of the Second International Conference on Boundaries in Ibero-America held at San José, Costa Rica, 14–17 November 1990. These papers, many of which have been translated from Spanish, seemed to complement the IBRU conference papers so well that it was decided to ask Dr Pascal Girot, who is co-ordinator of a major project on border regions in Central America based at CSUCA

(The Confederation of Central American Universities) to edit them for inclusion in the series. Volume 4 is thus symbolic of the practical cooperation which IBRU is developing with a number of institutions overseas whose objectives are much the same as IBRU's. The titles in the *World Boundaries* series are:

Volume 1 *Global Boundaries*
Volume 2 *The Middle East and North Africa*
Volume 3 *Eurasia*
Volume 4 *The Americas*
Volume 5 *Maritime Boundaries*

The papers presented at the 1991 IBRU conference in Durham were not specifically commissioned with a five-volume series in mind. The papers have been arranged in this way for the convenience of those who are most concerned with specific regions or themes in international boundary studies. Nevertheless, the editors wish to stress the importance of seeing the collection of papers as a whole: together, they demonstrate the ongoing importance of research into international boundaries on land and sea, how they are delimited, how they can be made to function peacefully, and perhaps, above all, how they change through time. If there is a single message from this impressive collection of papers it is perhaps that boundary and territorial changes are to be expected, and that there are many ways of managing these changes without resort to violence. Gatherings of specialists such as those at Durham in July 1991 and San José in November 1990 can contribute much to our understanding of the means available for the peaceful management of international boundaries. We commend these volumes as worthy of serious attention not just by the growing international community of border scholars, but also by decision-makers who have the power to choose between patient negotiation and conflict over questions of territorial delimitation.

Gerald H. Blake
Director, IBRU
Durham, January 1993

ACKNOWLEDGEMENTS

Much of the initial work on these proceedings was undertaken by IBRU's executive officer Carl Grundy-Warr before his appointment to the National University of Singapore early in 1992. It has taken a team of editors to complete the task he began so well. Elizabeth Pearson and Margaret Bell assisted in the preparation of the manuscripts for several of these volumes, and we acknowledge their considerable contribution. John Dewdney came to our rescue when difficult editorial work had to be done. In addition many people assisted with the organization of the 1991 conference, especially my colleagues Carl Grundy-Warr, Greg Englefield, Clive Schofield, Ewan Anderson, William Hildesley, Michael Ridge, Chng Kin Noi and Yongqiang Zong. Their hard work is gratefully acknowledged. We are most grateful to Tristan Palmer and his colleagues at Routledge for their patience and assistance in publishing these proceedings, and to Arthur Corner and his colleagues in the Cartography Unit, Department of Geography, University of Durham, for redrawing most of the maps.

Gerald H. Blake
Director, IBRU

INTRODUCTION

Carl Grundy-Warr

The contributions to this volume reflect the fact that border and border-lands studies are a multi-disciplinary field with a refreshingly high degree of cross-fertilization between disciplines. One of the aims of the International Boundaries Research Unit at the University of Durham, which sponsored the conference that made this volume possible, is to promote cross-disciplinary understanding of border and borderlands problems. Scholarship is well established on issues such as the evolution, delimitation, demarcation, functions and consequences of both land and maritime boundaries (Prescott 1987; Johnston 1987). In addition, there is a growing body of research, not on boundaries *per se*, but on borderlands as zones of contact, continuity and change; as zones of economic, political, social and cultural interaction (House 1982; Martinez 1986; Asiwaju and Adeniyi 1989; Grundy-Warr 1993). Furthermore, there is a strong concern with issues of conflict management and boundary-dispute resolution (Grundy-Warr 1990), on how to promote better transboundary co-operation and understanding (Martinez 1986; Asiwaju 1985) and on territorial alternatives to boundaries (Blake 1991). No volume covering *Eurasia* could hope to be comprehensive. Nevertheless, this volume does cover a geographically and conceptually broad scope. The purpose of this Introduction is to highlight key concepts and themes and to try to draw out certain common threads from the case studies. References to numerous other studies are used for readers who wish to explore some of these topics further and in far greater depth than is possible here.

The first chapter in the volume by historian Jeremy Black examines the significance and meaning of 'frontiers' in *ancien régime* Europe. His analysis of the evolution of mapping techniques, political maps and the 'pre-modern' conceptions of frontiers illustrate the importance of taking a historical-geographical perspective of the development of the world political map. Whilst frontiers in Europe were still a cause of dispute between polities in the period from the fourteenth to the eighteenth

1

centuries, the concept of frontiers was more legal/feudal than spatial. The politicization of frontiers and the development of linear conceptions of frontiers (or, rather, boundary lines as opposed to frontier zones) increased in the eighteenth century with technological advances in surveying and mapping, with the existence of more assertive states and state bureaucracies, and due to the impacts the French Revolution was to have upon the territorial world of much of Western Europe.

Another aspect of Black's study is that many different historical events and processes of change were at work in the partitioning of Europe and in the formation of new ideas about maps, territory and the political organization of space. Unfortunately, there are relatively few studies of this kind dealing with the period prior to the modern conception of political boundaries, although Gottmann (1973), a political geographer, has provided a valuable macro-level analysis of the significance of territory from classical times to the twentieth century. Black's chapter also suggests that we should be wary of simplified, unilinear explanations of changes to the political map. Sahlins (1989) also stressed this in his insightful examination of the making of the France–Spain boundary and the 'molecular' history of Catalan rural society in the Cerdanya borderlands in the eastern Pyrenees. His conclusions (1989: 298) about the Cerdanya are common for other borderlands, that 'the development of national identities and the formation of a territorial boundary line ... were historically variable and hardly unilinear processes'. Furthermore, whilst boundaries and borderlands have replaced vague frontiers or areas of overlapping jurisdictions similar to those existing in many parts of *ancien régime* Europe, they are still subject to changing relations between states over time. Understanding past processes of change and adaptation to change in borderlands helps to provide a clearer perspective on present-day conflicts, problems and changes (Martinez 1988; Minghi 1991).

One of the most sudden and dramatic historical changes to the world political map of recent times was the physical reunification of Germany. In Chapter 2, Hanns Buchholz examines some past and present problems related to the inner-German or German–German border. In particular, he discusses some of the problems the existence and disappearance of the boundary have created for spatial economic planning. The boundary produced very different 'regional' problems on either side. In the German Democratic Republic to the East the state-planners went to enormous expense to try to hermetically seal the boundary and they even attempted to ensure that a 'politically correct' population inhabited the zone nearest to the fortified border. In the Federal Republic of Germany to the West the authorities' main concern was not

2

so much with managing the boundary as with how to cope with large areas of the borderland that had effectively become economically peripheral to the main centres of economic activity. Reunification means that some parts of the old German–German borderlands may become new dynamic centres within Germany and Europe. Nevertheless, borderlands segmentation means that there will still be sections of the borderlands that remain economically backward, just as there were prior to 1945 (Ante 1991). Buchholz also points to new legal-economic problems that have developed on each side of the old boundary affecting land-use, such as 'cross-border' land, as well as a dilapidated transboundary infrastructure virtually untouched since the 1940s.

Reunification of the economy and society of Germany will be a very lengthy and difficult process. Buchholz suggests that people's 'mental maps' have not necessarily adapted to the new political realities. Walls, fences and watch-towers are far easier to remove than psychological boundaries manifested by divergent values, perceptions and aspirations fostered by living in two politically and economically opposed systems. Buchholz's chapter suggests that future research on the old German–German borderlands is necessary in order to see how new political, social and economic realities and how daily interaction between formerly divided Germans are changing the human landscape and spatial planning within 'the heart' of Germany. This may also help us to appreciate how the attitudes and mental maps of the borderlanders themselves are changing.

The removal of human-made divisions such as the German–German boundary is a rare event in the twentieth century. It is symbolic of the dismantling of the so-called 'Iron Curtain' between communist states to the East and capitalist states to the West of Europe. This geopolitical over-simplification of the political map of Europe tended to dominate foreign policy during the Cold War years. The geopolitical realities of Europe in the 1990s are more complex, as indicated by Michel Foucher in Chapter 3. According to Foucher, the fundamental geopolitical contradiction of Eastern Europe is that it comprises fewer states (thirty-three) than nations or ethno-linguistic entities with a national vocation (around fifty). For more than half a century this situation has remained unchallenged and the Helsinki Conference on Security and Co-operation in Europe in 1975 confirmed the principle of fixed borders. With the end of Soviet military control over much of Europe and with German reunification, this territorial order is in question. Borders may be and *are* being modified once more. The question remains how the right of self-determination of people is to be reconciled with the principle favouring the stability of boundaries (Higgins 1992).

Foucher studies the principal fault lines in modern Europe. With the dismantling of the ideological and military division along the 'Iron Curtain', the 'new Europe' is undergoing reconstruction along a new rift line: on one side Western and central Europe and, on the other, an Eastern Europe in crisis. This unstated line ominously coincides with one of the oldest sociocultural frontiers of Europe, that which divides the societies based on Western Christian civilization (Catholic and Protestant) and those based on the Orthodox religion (and, in south-eastern Europe, the legacy of the Ottoman Empire). While the first can be affiliated to the European Community (EC), the second, with a less-certain internal situation, may be left to fate.

The question is to know whether the 're-composition' of the continent represents an element of innovation, or on the contrary, a rehash of old geopolitical scenarios. Michel Foucher's analysis is based on very detailed research carried out by l'Observatoire Européen de Géopolitique at Lyon, which has produced a wealth of data on the historical evolution of the political, economic, social, cultural and ethnic maps of Europe, and adds to our understanding of the existing political map and the contradictions underlying it. The resurgence of ethno-nationalism and bloodshed in the Balkans (see Poulton 1991) has come as a shock to the states of Western Europe at a time when they are busily diluting their national border functions for an open, single European market. It is not surprising to find that the ancient religious and socio-cultural fault lines Foucher describes cut right through the territory of the old Yugoslavia. One of the problems faced by the peace-makers from the EC and the United Nations is that ancient divisions, remembered or perceived nationalist history, and old scores are partially fuelling the current conflict. Furthermore, it is also apparent that new boundaries 'based largely on crude nationalism and raw military strength' are unlikely to lead to a stable peace in this region for years to come (Englefield 1992).

An extension of the Christian–Muslim fault line that cuts through the Balkans lies in the eastern Mediterranean island of Cyprus, which has also suffered from bitter intercommunal and ethno-territorial conflict. Indeed, the United Nations (UN) has had nearly thirty years of involvement in the Cyprus conflict between the Greek and Turkish Cypriots. Chapter 4 by Carl Grundy-Warr focuses on the role of the UN Force in Cyprus (UNFICYP) in two distinct periods. First, the decade prior to *de facto* partition of the island into two almost mono-ethnic territories is used to illustrate the UN role in relation to a process of patchwork territorial segregation of the two main communities. Second, in the period since partition in 1974 UNFICYP's primary roles have been in relation to the monitoring and management of a demilitarized zone between the front lines of the Turkish Army and the Cyprus National Guard made up of Greek Cypriots. UNFICYP has had to cope with two complete transformations to the political and human landscape of Cyprus. It is suggested that there are important lessons to be learnt regarding the potentialities and limitations of third-party conflict management from the long-standing UN mission in Cyprus.

One of the important issues addressed by this chapter is whether or not the presence of the UN peacekeepers has actually contributed to the political stalemate and hardened the physical division of the island. This chapter suggests that whilst UNFICYP has tried to maintain an unsatisfactory military *status quo* as part of its efforts to monitor the cease-fire and prevent a recurrence of fighting, it has also tried to minimize the hardships created by an arbitrary division of the island's space. A sudden withdrawal of the Force would make violent conflict more likely and would probably lessen the chances of a peaceful settlement. Whatever the future holds, the *de facto* partition of Cyprus, like the contentious borders in the Balkans, represents a virtually impenetrable physical barrier to daily intercommunal relations between ordinary people and a source of continuing mistrust between two communities who have shared the same territory in the past. Unlike Germany, a political solution is more likely to be on some sort of bizonal, bicommunal federal basis. Nevertheless, any political settlement will of necessity be complex to achieve on the ground. There are too many scars for quick solutions to heal. A solution would have to involve compromises from both sides at every level of Cypriot society. One of the facts of the German reunification has been that human-made physical barriers can be quickly broken down, but the phobias and mental walls created by years of a separate territorial existence are much harder to remove.

In Chapter 5 Julian Minghi examines examples of permeable borders where increasing co-operation is creating new challenges. He builds on his earlier work (1963, 1981, Rumley and Minghi 1991) to examine the regions created by boundaries (the borderlands). In Western Europe there has been considerable progress in inter-state relations and in fostering cultural, economic and social interaction within borderlands. There are extremely well-developed transborder agencies, such as those of Regio Basiliensis, covering a trinational region overlapping France, Germany and Switzerland (Briner 1986; Gallusser 1991). Minghi identifies elements in the process of change in the human geography of borderlands which indicate that policies of growing harmony and co-operation between neighbours are reflected directly in the process of landscape change. Indeed, in many cases the borderlands have become the actual symbols and instruments of improved relations between states. Nevertheless, new spatial-political conflicts have evolved as neighbouring state planners look to their borderlands to develop inter-state level co-operation, particularly conflicts of interest between central government authorities and the interests of the borderland commu-nities. It is important to note that sub-national and local-level author-ities within borderlands are frequently engaging in 'international' diplomacy with similar agencies across the border (Duchacek 1986, 1988). Thus the study of borderlands entails an understanding of inter-actions at different spatial scales within and across national boundaries. Finally, Minghi speculates about the process of future change in the borderlands of the 'new Europe'. Of particular interest will be the borderlands overlapping the boundaries of the former front-line states of the old 'Iron Curtain'. It is likely that some of these borderlands may continue to be 'alienated' due to increasing border restrictions put up by EC members against large-scale immigration from non-EC countries. As Minghi concludes, the borderlands of Europe are likely to play a central role in the lengthy period of adjustment facing the 'new' Europe.

The first of the 'Asian' chapters is by Nurul Islam Nazem who high-lights a major problem faced by many neighbouring states in the developing world, that of hydropolitics. In this case the water resources across the India–Bangladesh boundary are discussed. The chapter focuses on the problems faced by the militarily and politically weaker state, Bangladesh. Rivers are a significant feature of the country's physical and cultural landscape. Proper management of these rivers is vital to its national development; but the utilization and management of these river courses are constrained by external factors, particularly those resulting from Indian policies which have had detrimental consequences on Bangladesh's water supply and environment. The author discusses some of the adverse effects of India's innumerable structural activities on the shared international waters with Bangladesh: as the lower-riparian state, Bangladesh is in a disadvantaged position vis-à-vis her large neighbour with regard to upstream developments. According to Nazem, India's policies are shaped with her regional geo-strategic goals in mind, which has militated against the development of a more rational, less wasteful transboundary water-management regime.

Whilst the India–Bangladesh case has many unique features, the use of water as a political weapon is common to many other cases, particu-larly between neighbours that have a history of mutual distrust (Anderson 1987; Naff and Matson 1984). It is not always politics that

6

leads to mismanagement of resources between states. Sometimes well-intentioned unilateral policies by upper-riparian states have unintended and devastating consequences for lower-riparian states, as may be partially the case here. The fact is that complex hydrological systems necessitate multi-use, basin-wide, systematic approaches to problems (Naff 1992). This in turn means the establishment of cross-border legal and administrative frameworks to deal with disputes and management problems of mutual concern to neighbouring states (Mumme 1992; Szekely 1992). In the case of some of the water-management issues between India and Bangladesh and the ever-present danger of huge socio-economic costs to local populations, effective transboundary environmental diplomacy (Carroll 1988) is urgently needed.

In Chapter 7 Dennis Rumley attempts to develop a framework for understanding conflicts in international border regions of the developing world. It is suggested that one of the principal causes of conflict in any political system is the failure to maximize the primary functions of politically organized space – participation, representation and resource allocation. Rumley argues that political geographers are uniquely placed to undertake evaluations of the political organization of space at any scale – national, regional, local – and in different developmental contexts in order to lend insights into such conflicts. Such a perspective inevitably assumes a 'reformist' philosophical orientation which possesses policy implications.

This perspective underpins a consideration of conflict in the Thai–Malaysian border region. Five main approaches to the analysis of the causes of border-region conflict were evaluated (positivist, political development, colonial-boundary, ethno-religious nationalism, core–periphery) and it is suggested that no one approach possesses theoretical or practical pre-eminence. Rumley argues that expressions of conflict in the region can best be understood from the point of view of a combination of factors arising out of the impact of European colonialism, the structure and policies of contiguous states and the aspirations of the regional populations. Furthermore, a comprehensive understanding of the nature of border conflict must also take into account global social and political changes as well as regional cultural norms and practices.

Rumley's approach may have wider application in other borderlands of South East Asia, where conflict rather than co-operation is more common (Lintner 1991; Lee 1982; Lim and Vani 1984; May 1991; Grundy-Warr 1993). His stress on internal structures is necessary because most of the conflicts in borderlands involve state–minority relations, civil war and ethno-territorial conflict. As such, they are very different from disputes about boundary lines mostly analysed at the

7

inter-state level (Prescott 1987). It is also important to stress that Eurocentric or Anglo-American concepts of boundaries and boundary evolution do not fit the Asian context (Lee 1982). As Lim (1984: 69) observed, traditional 'Asian' concepts of frontiers based on loosely interpenetrating political systems and on the notion 'that boundaries are never rigid nor once and for all time' are still useful in helping us to understand certain aspects of contemporary inter-state and intra-state politics. Rumley's work should provide a framework for other researchers interested in borderlands conflicts to build on.

Some parts of Asia are currently experiencing extremely rapid economic transformation and economic growth. The Hong Kong–South China borderlands (Chapter 8) is perhaps one of the most economically dynamic transboundary regions in the world. Some economists (such as Chia and Lee 1992) have already described the South China–Hong Kong area as an example of a 'sub-regional economic zone' (SREZ). The SREZs are characterized by substantial cross-border investments or industrial projects involving at least two national economies or parts of those economies, particularly border areas. Other examples of SREZs include the so-called 'growth triangle' involving Singapore, Johor state (Malaysia) and the Riau Province (Indonesia), and the ambitious Tumen River project involving East Russia, China, Mongolia and North and South Korea. Nevertheless, whilst we may use broad descriptive labels we should be cautious about making too many generalizations about very distinct cases of transboundary economic co-operation. Each example has a unique historical socio-economic and political context requiring both macro- and micro-level research.

The case of the Hong Kong–China borderlands is unique in many ways, not least because it has been a contact-point between communist China and the outside world. Its boundary has separated a British-administered capitalist city from Chinese socialism. Yet, even in China's most isolationist years, the boundary remained open to some degree of trade and contact between the two places. China's economic reform programme, which began in 1978, gave major roles to external trade and investment. Since then Hong Kong has expanded vastly its role as an economic frontier where China and capitalist countries could meet and interact. Patricia Goodstadt argues that the two most significant changes in boundary relations since the establishment of China's 'open door' economic policies were the closure of the boundary to illegal immigration into Hong Kong and the opening of the Chinese side to investment and exports from Hong Kong. The former had important implications for the Hong Kong labour market, which tended to act as a 'push' factor persuading Hong Kong-based manufacturers to relocate and invest in assembly-type operations in China, particularly in Guang-dong Province and the Shenzhen Special Economic Zone.

To understand the economic and social changes under way in the Hong Kong–China borderlands more research is required into the cross-border social and kinship ties that exist and do have an influence on the locational decision-making of businesses (Smart and Smart 1991). The bulk of transboundary investments are from Hong Kong (Henderson 1991) and comprise many small and medium-sized Hong Kong-based or overseas Chinese companies. For instance, Chi Kin Leung (1993) has provided a useful study of the significance of preexisting kinship and business ties on Hong Kong's production subcontracting in the Zhujiang or Zhi Jian (Pearl River) Delta region. He found that these linkages help to insure exchange reliability and facilitate further cross-border inter-firm collaboration and transactions. Building an explana-tory framework to understand changes in the borderland, both currently and in the period after the reintegration of Hong Kong into China, will also require a better understanding of the development of new economic 'strata' in the commercialized territories of China, including officials dealing with foreign capitalists, private entrepreneurs and the large group of People's Republic Chinese who live and work in Hong Kong and abroad (Sklair 1991).

In 1997 the international boundary will disappear. This has led to much speculation concerning the future of Hong Kong and the South-China border zone which forms Hong Kong's hinterland and whose fortunes are increasingly tied to those of the current British colony. It has been suggested that in the absence of rapid democratic reforms Hong Kong's future viability as a capitalist enclave is in doubt (Henderson 1991). Another perspective is that Hong Kong will simply be incorporated into a much larger geographic area of economic growth. Xiangxing Lu (1992) argued that the Pearl River Delta, with three international metropolitan areas – Hong Kong, Guangzhou (Canton) and Macao, plus numerous small and medium-sized settle-ments – is destined to become a leading global economic centre. Good-stadt argues that China's management of its boundary relations with Hong Kong reveals elements of how Beijing expects the 'one country, two systems' formula to operate. Hong Kong would continue to be a business centre partially insulated from China's communist system (two systems) and yet would function as part of China's national system through its dependence on the rest of the country for imports and for manufactured capacity (one country).

As Patricia Goodstadt suggests, it would be virtually impossible, once the boundary has gone, to insulate Hong Kong without offering similar advantages to extensive areas of Guangdong Province, especially the neighbouring territory of the Shenzhen Special Economic Zone. This has led some writers to suggest that the northern boundary of the Shenzhen zone will become a significant economic border for Hong Kong (Kelley 1990). During the 1980s Shenzhen's northern border did act as a significant control on the flow of exports and imports, foreign currency and labour (Sklair 1991). Clearly, there is a need for long-term research agendas on the Guangdong–Shenzhen–Hong Kong territory before and after 1997.

Kimie Hara's chapter (completed in February 1992) on the Northern Territories (Kuriles) territorial issue between Japan and Russia discusses the various dimensions (strategic, economic, geographical, historical-political) of the problem. Her analysis is particularly strong on the period of *rapprochement* from Gorbachev's 'new-thinking diplomacy' to the first few months of the period after the Soviet Union disintegrated, when the signs looked good for some kind of political compromise on the territorial issue largely as a result of Russia's dire economic condition and the need for economic and financial support from the international community, including Japan. Whilst Hara remains optimistic that the removal of post-Cold War impediments to improved bilateral relations provides a basis for long-term historic political change on the Northern Territories issue, there are significant obstacles to overcome, particularly the great political and socio-economic uncertainties within the new Russian Republic.

Within Russia there was a desire amongst the democrats who had gathered around President Boris Yeltsin to prove Russia's credibility as a partner of the industrialized and developed world in order to create more solid foundations for democracy. Agreement with Japan, plus the influx of large-scale Japanese investments, would be proof of that credibility. Nevertheless, as Buszynski (1993) has pointed out, there have been three strong reasons preventing Moscow from making the necessary territorial concessions which Tokyo has demanded as a precondition to direct economic aid to Russia: first, the possibility that compromise on the Northern Territories would lead to a 'chain-reaction' of territorial claims against Russia, of which there are many (Kolossov 1992); second, the strategic maritime interests have not disappeared, although they are diminished with the end of the Cold War; third, the Governor of Sakhalin, Valentin Fedorov, has campaigned that the islands represent 'ancient Russian territory' and that the rights of the estimated 34,000 Russian residents who inhabit the islands, including around 7,000 military personnel, should not be forgotten. As Buszynski (1993: 51) put it, Fedorov's defence of Russian settlers 'struck a deep chord among Russians who feel anxious about the fate of an estimated 25 million of their compatriots, who now find themselves outside the borders of Russia'. Another powerful indication of the domestic political troubles Yeltsin has faced was the cancellation of his much-publicized proposed visit to Tokyo in September 1992 which came shortly after a statement by the Russian Parliament that it would not ratify any decision made by the President over the future status of the disputed islands.

The Russo-Japanese case highlights the intricacies of territorial and border disputes in general. The fact is that such disputes often reflect not only the state of inter-state relations but often domestic political issues as well. Furthermore, in situations of economic crisis such as the on-going situation in Russia, it is easier for nationalists to manipulate public opinion and to attach their own historical interpretations and meanings to territorial issues. Under such circumstances it is very difficult to forge the necessary domestic compromises needed to begin a peaceful international-negotiation process.

If bilateral disputes can at times be deadlocked because of dia-metrically opposed national interests, then the 'tangled web' of national interests (China, Vietnam, Taiwan, the Philippines, Malaysia, and also Brunei) involved in the Spratlys dispute represents a huge conflict-reso-lution problem. This is also because the Spratlys dispute is multi-dimen-sional (historical, strategic, geographical, economic, political and legal; Thomas 1990; Valencia 1989; Hill, Owen and Roberts 1991). In Chapter 10 Peter Kien-hong Yu discusses the position of the Republic of China (ROC) (Taiwan), and examines the military, political and econ-omic aspects of Itu Aba Island's status in the Spratlys dispute. From his discussions with officials of the National Defence University and State Oceanic Administration in the People's Republic of China, Yu argues that the island will remain in the ROC's hands in the foreseeable future. As Yu indicates, the ROC is unlikely to renounce Chinese 'sovereignty' over the Spratlys, although Taipei has, in his words, called for 'a peaceful resolution to the dispute over sovereignty through bilateral negotiations'.

By focusing on the position of one of the claimants Yu's chapter clearly illustrates some of the key problems facing any international efforts to resolve or even to manage the conflict over the Spratlys. As Thomas (1990: 425) has pointed out, 'bilateral agreements for peaceful settlements or setting aside of the issue do not necessarily indicate willingness to ultimately yield territory to the opposite party'. Both the People's Republic of China and Vietnam have all-inclusive claims to the Spratlys and have indicated that they desire peaceful solutions, but the extent to which they are prepared to compromise on the territorial issue is in considerable doubt. It is true that the geopolitics of the South China Sea has changed in the aftermath of the Cold War and due to the increasing importance of 'geo-economic' strategic considerations, such as transnational investments, financial and trading links, particularly with the opening-up of the economies of China and Vietnam (Polomka 1991). Even so, the actual 'settlement' of the Spratlys dispute is still a long way off, given the 'national sentiments involved' and 'the variety of interpretations and criteria suggested in conventional and customary international law'; rather, it is suggested that in place of utilizing the rules of delimitation and boundary-making it may be more useful to rely on 'the principle of regional co-operation in semi-enclosed seas stated at UNCLOS III' (Rhee and MacAulay 1988: 95–6, 99). A start may well be to encourage through either sub-regional regimes, such as the Association of South-East Asian Nations (ASEAN) or through regional regimes, such as the Asia Pacific Economic Cooperation (APEC) group, or even a 'Spratlys Authority', a basis for co-operation between the littoral and claimant states to preserve the marine environment and to conduct scientific research. Such a functionalist approach to the dispute may be the best way forward. As indicated in Yu's chapter, the ROC is willing to consider co-operation in technical and environmental fields as well as the 'joint exploitation' of resources.

This brings us to the final chapter, written by the cross-disciplinary team of Auburn, Forbes and Scott, comparing different joint-development regimes involving respectively, Thailand and Malaysia; Japan and South Korea; Antarctica, and Indonesia and Australia. The treaties cover more than just exploration and exploitation of hydrocarbon resources in the substratum of the sea. For example, the Japan/South Korea Agreement provides for the adjustment of fisheries interests. The Timor Gap Treaty extends to air-traffic services co-operation. The Convention on Antarctica also provided detailed and binding environmental principles, whilst the Timor Gap Treaty delegates the environmental issues to the Joint Authority. There is also a close relationship between the functional demands and the contents of the agreements. Thus the Thailand/Malaysian Memorandum deals with a small geographical area, with few oil prospects, is brief and covers significant issues in a generalized manner. By contrast, the Timor Gap Treaty covers the highly prospective Kelp Structure and attempts to deal with a broad range of issues in some detail.

The authors point out that all the agreements, with the exception of CRAMRA, were drafted as interim agreements. Whether there is an undertaking to attempt to negotiate permanent boundaries or not, the other regimes are likely to become much more long-standing arrangements. According to Blake (1992) there are now approximately eighteen agreements in which shared maritime space of one kind or another has been delimited, involving approximately twenty-nine states. Furthermore, there is the possibility of new shared regimes as the advantages of the existing joint-development zones are more widely appreciated. It has been suggested that shared territorial arrangements and shared zones of restricted economic activity could provide temporary relief to potential violent conflict on land. At least shared territorial arrangements deserve more attention as ways to manage conflicts within or between states (Blake 1991), in a similar way perhaps but as a functional extension of existing examples of United Nations-supervised demilitarized zones. If states are reluctant to concede territory or cannot agree over precise boundaries, then it may be that shared sovereignty over specific disputed geographical areas can become part of more lasting conflict-settlement provisions.

Most of the contributions in this volume are in some way concerned with conflicts over borders or territory. This tends to give a misleading picture of the current world political map, which is characterized more by peaceful boundaries than troublesome ones; but even where there are no inter-state disputes over boundaries there are often many human problems simply because of the existence of artificial political lines on maps (Grundy-Warr and Schofield 1990). Although we are still a very long way from 'the borderless world' envisaged by Kenichi Ohmae (1992), global finance capital (Thrift 1992), the transnationalization of production and services world-wide (Dicken 1992) and the influence of world media networks (Thompson 1990) are tending to reduce the significance of national boundaries. As Thrift put it, we need to 'look beyond the nation-state as the traditional unit of analysis'. This may be true, but we can also not ignore the continued potency of territorial and border issues and the contradictions between the rights of states and the rights of people to self-determination, which has led to the break-up of states and the creation of many new political boundaries (Higgins 1992; Joffé 1992). Furthermore, the artificial division of natural resources by political borders makes the study of transborder relations, processes and inter-state resource management mechanisms a critical research field for helping to find effective solutions to many highly sensitive international problems. Finally, there is a growing body of literature that focuses on international borderlands as zones of considerable socio-economic interaction blurring the edges of nation-states, and as zones where there is often an intermingling of different cultures, and where the destinies of neighbouring peoples merge, sometimes in conflict with the interests of central state authorities. Such research adds much to our understanding of our politically divided world. It is hoped that this volume will stimulate further interest and generate more research on both boundary-related issues and on borderlands.

REFERENCES

Anderson, E.W. (1987) 'Water Resources and Boundaries in the Middle East', in G.H. Blake and R.N. Schofield (eds), *Boundaries and State Territory in the Middle East and North Africa*, Cambridge: Menas Press, 85–98.

Ante, U. (1991) 'Some Developing and Current Problems of the Eastern Border Landscape of the Federal Republic of Germany: The Bavarian Example', in D. Rumley and J.V. Minghi (eds), *The Geography of Border Landscapes*, London: Routledge, 63–85.

Asiwaju, A.I. (1985) *Partitioned Africans. Ethnic Relations Across Africa's International Boundaries, 1884–1984*, London: C. Hurst & Company.

Asiwaju, A.I. and Adeniyi, P.O. (eds) (1989) *Borderlands in Africa. A Multi-Disciplinary and Comparative Focus on Nigeria and West Africa*, Lagos: University of Lagos Press.

Blake, G.H. (1991) 'Shared Zones as a Solution to Problems of Territorial Sovereignty in the Gulf States', The Territorial Foundation of the Gulf States: Geopolitics and International Boundaries Conference, Geopolitics Research Centre, School of Oriental and African Studies, London.

―――― (1992) 'Territorial Alternatives', *Boundary Bulletin*, 3: 9–12.

Briner, H.J. (1986) 'Regional Planning and Transfrontier Cooperation: The Regio Basiliensis', in O.J. Martinez (ed.), *Across Boundaries. Transborder Interaction in Comparative Perspective*, El Paso: Texas Western Press, 45–56.

Buszynski, L. (1993) 'Russia and Japan: The Unmaking of a Territorial Settlement', *The World Today*, 49/3: 50–4.

Carroll, J.E. (ed.) (1988) *International Environmental Diplomacy*, Cambridge: Cambridge University Press.

Chia, S.Y. and Lee, Tsao Y. (1992) 'Subregional Economic Zones: A New Motive Force in Asia–Pacific Development', 20th Pacific Trade and Development Conference, Washington DC.

Dicken, P. (1992) *Global Shift. The Internationalization of Economic Activity*, London: Paul Chapman Publishing.

Duchacek, I.D. (1986) 'International Competence of Subnational Governments: Borderlands and Beyond', in O.J. Martinez (ed.), *Across Boundaries. Transborder Interaction in Comparative Perspective*, El Paso: Texas Western Press, 11–30.

―――― (1988) *Perforated Sovereignties and International Relations: Trans-Sovereign Contacts of Subnational Governments*, New York: Greenwood Press.

Englefield, G.E. (1992) 'Yugoslavia, Croatia, Slovenia: Re-Emerging Boundaries', *Territory Briefing*, 3.

Gallusser, W.A. (1991) 'Geographical Investigations in Boundary Areas of the Basle Region ("Regio")', in D. Rumley and J.V. Minghi (eds), *The*

Geography of Border Landscapes, London: Routledge, 31–42.
Gottmann, J. (1973) *The Significance of Territory*, Charlottesville: The University Press of Virginia.
Grundy-Warr, C. (ed.) (1990) *International Boundaries and Boundary Conflict Resolution*, Durham: Boundaries Research Press.
—— (1993) 'Coexistant Borderlands and Intra-State Conflicts in Mainland Southeast Asia', *Singapore Journal of Tropical Geography*, 14/1, 42–57.
Grundy-Warr, C. and Schofield, R.N. (1990) 'Man-made Lines that Divide the World', *Geographical Magazine*, LXII/6: 10–15.
Henderson, J. (1991) 'Urbanization in the Hong Kong–South China Region: An Introduction to Dynamics and Dilemmas', *International Journal of Urban and Regional Research*, 15/2: 169–79.
Herzog, L.A. (1991) *Where North Meets South: Cities, Space and Politics on the U.S.–Mexico Border*, Austin: Center for Mexican American Studies, University of Texas.
Higgins, R. (1992) 'Keynote Address. Panel on Principles of Boundary Delimitation', International Boundaries: Political, Legal and Strategic Implications Conference, London.
Hill, R.D., Owen, N.G. and Roberts, E.V. (eds) (1991) *Fishing in Troubled Waters*, Hong Kong: Centre for Asian Studies.
House, J.W. (1986) *Frontier on the Rio Grande. A Political Geography of Development and Social Deprivation*, Oxford: Clarendon Press.
Johnston, D.M. (1987) *The Theory and History of Ocean Boundary Making*, McGill-Queen's University Press.
Kelley, I. (1990) 'The Hong Kong–China Boundary', *Boundary Briefing*, 1.
Kolossov, V.A. (1992) 'Ethno-Territorial Conflicts and Boundaries in the Former Soviet Union', *Territory Briefing*, 2.
Lee, Y.L. (1982) *Southeast Asia: Essays in Political Geography*, Singapore: Singapore University Press.
Leung, C.K. (1993) 'Personal Contacts, Subcontracting Linkages, and Development, in the Hong Kong–Zhujiang Delta Region', Annals of the Association of American Geographers, 83/2, 272–302.
Lim, J-J. (1984) *Territorial Power Domains, Southeast Asia, and China*, Singapore: Institute of South-East Asian Studies.
Lim, J-J. and Vani, S. (eds) (1984) *Armed Separatism in Southeast Asia*, Singapore: Institute of South-East Asian Studies.
Lintner, B. (1991) 'Cross-Border Drug Trade in the Golden Triangle (S.E. Asia)', *Territory Briefing*, 1.
Lu X.X. (1992) 'Changing Patterns of Regional Economic Development in Post-Reform China', 27th Congress of the International Geographical Union and Assembly, Washington DC.
Martinez, O.J. (ed.) (1986) *Across Boundaries. Transborder Interaction in Comparative Perspective*, El Paso: Texas Western Press.
—— (1988) *Troublesome Border*, Tucson: The University of Arizona Press.
May, R.J. (1991) 'The Indonesia–Papua New Guinea Border Landscape', in D. Rumley and J.V. Minghi (eds), *The Geography of Border Landscapes*, London: Routledge, 152–68.
Minghi, J.V. (1963) 'Boundary Studies in Political Geography', *Annals of the Association of American Geographers*, 53: 407–27.

—— (1981) 'The Franco-Italian Borderland: Sovereignty Change and Contemporary Developments in the Alpes Maritimes', *Regio Basiliensis*, 22: 232–46.

Mumme, S.P. (1992) 'New Directions in United States–Mexican Transboundary Environmental Management: A Critique of Current Proposals', *National Resources Journal*, 32: 539–62.

Naff, T. (1992) 'The Litani in the Context of Scarce Water Resources in the Middle East', Conference on Peace-Keeping, Water and Security in South Lebanon, London.

Naff, T. and Matson, R.C. (1984) *Water in the Middle East. Conflict or Cooperation?*, Boulder: Westview Press.

Ohmae, K. (1990) *The Borderless World. Power and Strategy in the Interlinked Economy*, USA: Harper Business.

Polomka, P. (1991) 'Strategic Stability and the South China Sea: Beyond Geopolitics', in R.D. Hill, N.G. Owen and E.V. Roberts (eds), *Fishing in Troubled Waters*, Hong Kong: Centre for Asian Studies, 36–47.

Poulton, H. (1993) *The Balkans: Minorities and States in Conflict*, London: Minority Rights Group.

Prescott, J.R.V. (1987) *Political Frontiers and Boundaries*, London: Allen & Unwin.

Rhee, S-M. and MacAulay, J. (1988) 'Ocean Boundary Issues in East Asia: The Need for Practical Solutions', in D.M. Johnston and P.M. Sanders (eds), *Ocean Boundary Making: Regional Issues and Developments*, London: Croom Helm, 74–108.

Rumley, D. and Minghi, J.V. (eds) (1991) *The Geography of Border Landscapes*, London: Routledge.

Sahlins, P. (1989) *Boundaries: The Making of France and Spain in the Pyrenees*, Berkeley: University of California Press.

Sklair, L. (1991) 'Problems for Socialist Development: The Significance of Shenzhen Special Economic Zone for China's Open Door Development Strategy', *International Journal of Urban and Regional Research*, 15/2: 197–215.

Smart, J. and Smart, A. (1991) 'Personal Relations and Divergent Economies: A Case Study of Hong Kong Investment in South China', *International Journal of Urban and Regional Research*, 15/2: 216–33.

Szekely, A. (1992) 'Establishing a Region for Ecological Cooperation in North America', *Natural Resources Journal*, 32: 563–622.

Thomas, B. (1990) 'The Spratly Islands Imbroglio: A Tangled Web of Conflict', in C. Grundy-Warr (ed.), *International Boundaries and Boundary Conflict Resolution*, Durham: Boundaries Research Press, 413–28.

Thompson, J.B. (1990) *Ideology and Modern Culture*, Cambridge: Polity Press.

Thrift, N. (1992) 'Muddling Through: World Orders and Globalization', *The Professional Geographer*, 44/1: 3–6.

Valencia, M.J. (1989) 'Maritime Claims Bedevil the Spratly Islands: A Co-Operative Regime Could be the Solution', *Centerviews*, 7/3: 3.

Part I

EUROPE

1

BOUNDARIES AND CONFLICT

International relations in
ancien régime Europe

Jeremy Black

Our boundaries should be clearly, certainly, and circumstantially
defined, so that no future disputes may arise about them.
('Valerius Publicola' in *London Evening Post*, 11 September 1762)

This chapter addresses the interrelated questions of the part played by
frontier disputes in the diplomacy of *ancien régime* Europe, the
changing notions of frontier, as evidenced in the European diplomacy of
the period, and the role played by more effective mapping in changing
contemporary conceptions of frontiers and in the resolution of frontier
disputes.[1]

FRONTIERS IN EURASIA

Ancien régime European states faced three related but different
problems in defining frontiers. They can be classified as follows, without
the implication that this classification should be seen as too rigid. First,
there was the delineation of frontiers between European states, more
particularly in areas of long-standing settlement or at least control by a
Christian polity. Second, there was the delineation of frontiers between
European states in areas where there were no historic claims or long-
standing European settlements. As the Western hemisphere passed
under European control after 1492, this was especially important.
Third, there was the delineation of frontiers between European and
non-European societies. These non-European societies could appear
somewhat unsophisticated. There were differing notions of sovereignty
in non-European societies. Maps were of limited use as representations

19

of power in a territorial sense if the basis for the concept of ownership was neither legal nor territorial. Guillaume Delisle's *Carte d'Afrique* (Amsterdam, c. 1700) misleadingly divided the whole of Africa into kingdoms with clear frontiers (Freeman-Grenville 1991: 84, 72; Ade Ajayi and Crowder 1985). Maps were more useful if there was a territorial sense of ownership, a notion of fixed frontiers and a use of natural features as boundaries. Natural features were cited in treaties between the British North American colonies and Indian tribes. Treaties such as that of 18 November 1765 with the Lower Creeks in Florida helped to keep the peace (Crane 1929; Cumming 1962; DeVorsey Jr 1966: 149–57; Cockran 1967; Sosin 1967; Cashin 1992: 214–22, 229, 238–47).

Asiatic societies were certainly not politically unsophisticated. The Ottoman Empire had common boundaries with Venice, Austria, Poland and Russia, and the period witnessed the territorial meeting of Russia with both China and Persia, and their negotiation of border treaties: Nerchinsk with China in 1689 and treaties with Persia in 1723, 1729 (Rescht) and 1732 (Rescht again). The problem of defining land frontiers in Eurasia could be considerable. The Austrian and Turkish commissioners who sought to clarify their new frontier after the Peace of Carlowitz of 1699 faced the ambiguous and contradictory wording of the peace treaty on such matters as the 'straight' line of one portion of the frontier, the 'ancient' frontiers of Transylvania and the future status of islands where the frontier followed rivers. Following the Austro-Turkish war of 1737–9 and the subsequent Treaty of Belgrade there were lengthy negotiations to settle the new border, and a satisfactory settlement was not negotiated until 1744. There were serious Austro-Turkish differences over their Bosnian border in 1784–5.

The value of international frontiers on the Eurasian border was limited. A recent historical atlas of the Ukraine comments on the map of Ukrainian lands after 1569 that 'the international boundaries between the lower Dnieper and lower Donets' rivers as marked on this map were really only symbolic, because this whole region was a kind of no-man's land dominated by nomadic and free-booting communities of Zaprozhian Cossacks and Nogay Tatars' (Abou-El-Haj 1969; Stoye[2]; Olson 1975: 152; Magocsi 1985: map 10; *Gentleman's Magazine* 1791: 861).

Aside from the problem of defining major land frontiers in Eurasia, there was also the difficulty of determining the relationship between the coastal enclaves of European trading companies or states in southern and eastern Asia and the locally dominant Asiatic powers. Questions of sovereignty and jurisdictional relationship were prominent, but the situation was far from uniform. Aside from differences between the imperial organizations and pretensions of different European societies, it was also the case that some Asiatic polities, such as the Indian Mughal and Persian Safavid empires, provided only loose hegemonies in which it was possible for European interests to establish semi-independent territorial interests, akin to those of some Asiatic regimes (Bayly 1989: 46–7; Bonney 1971: 52–101; Bassett 1971: 73–80).

Natural boundaries were an obvious basis for the Eurasian land frontier between Europe and Asia. There was no jurisdictional definition of territory reflecting long-established political interests. Instead, force operated with scant reference to historical claims. The exact course of this frontier was most important in areas of settlement and, as these were riverine, it was rivers that provided the necessary definition. The Amur had marked the crucial section of the Russo-Chinese border between 1644 and 1689, but Russia lost the Amur region under Nerchinsk. For much of the eighteenth century the frontier between Russia and central Asia east of the Caspian – in so far as one can speak of one – followed the Ural and Irtysh rivers. The Terek and the Kuban defined much of Russia's frontier in the Caucasus in the late eighteenth century. Further west, the Dnieper, Bug, Dniester and Pruth marked successive stages of Russia's advance across transpontine Europe and towards the Balkans. Similarly, the Oltul, Muresul, Tisza, Danube and Sava played an important role in defining the Austro-Turkish border between 1699 and 1878. These frontiers became better mapped in the eighteenth century. Reviewing Lewis Evans's *Analysis of a General Map of the Middle British Colonies in America*, Dr Johnson wrote in 1756 that 'the last war between the Russians and the Turks [1736–9] made geographers acquainted with the situation and extent of many countries little known before' (*Literary Magazine*, 15 October 1756). This continued to be the case. Conflict encouraged military mapping and the commercial production of maps. The *Journal Politique de Bruxelles* of 2 February 1788 advertised a map of the northern and north-western littoral of the Black Sea that would help those interested in the recently commenced Russo-Turkish war to follow its course.

This chapter concentrates on the European frontiers of European powers, but it is worth dwelling on the Eurasian frontier because it poses in some respect a contrast, or, from another viewpoint, a different point on a continuum, from which inter-European disputes can be assessed. Power and pragmatism divorced from the feudal legacy of jurisdictional issues and non-linear frontiers were dominant on the Eurasian frontier. This was certainly the case with the Russian impact in the Balkans, the Caucasus and further east. It has been argued by Atkin that an obsessive craving to make original lines invulnerable by creating new ones further forward impelled the Russians forward in an atmosphere of mistrust, insecurity and sensitivity to their frontiers. In turn, this exacerbated the attitudes of others, accentuating the vortex of opportunity, distrust, opportunism and conflict (Lang 1957; Fisher 1970; Jewsbury 1976; Atkin 1980).

Although it was not part of the Eurasian frontier, there had been a similar uncertainty about the frontier between Russia and Sweden-Finland. It was not until a conflict ended by the Treaty of Teusina (1595) that a frontier was drawn for the first time between Finland and Karelia from the isthmus to the White Sea, Russian control of the Kola peninsula and Swedish control over most of Lapland being acknowledged. Nevertheless, Norwegian-Russian 'common districts' – areas of mixed taxation – remained, until partitioned in 1826. There were serious disputes over the frontier between Sweden and Denmark-Norway north of the Arctic Circle (Somme 1968: 15–17; Kirby 1990: 20).

FRONTIERS AND SETTLEMENTS

Power and pragmatism were also dominant in the case of transoceanic relations between the European powers. Territorial disputes outside Europe had played a role from the outset, but they became more acute as the frontier of European power increasingly disappeared in terms of territorial claims, although obviously not of settlements. Such disputes were most apparent in coastal regions – generally the only well-mapped areas and the ones that were most subject to exploitation. European knowledge of the interior of other continents was limited and they were thus poorly mapped. This also reflected the navigational rationale of many maps. For example, the Venetian Coronelli's *Route maritime de Brest à Siam et de Siam à Brest* (Paris, 1687) was essentially a map of coastal regions. Etienne de Flacourt's map of Madagascar (1666) was accurate largely for the south-east of the island, where the French had established Fort Dauphin in 1642. In D'Anville's *Carte de l'Inde* of 1752 most of east-central India was labelled 'Grand espace de pays dont on n'a point de connoissance particulière'. Desnos' map *L'Asie* (Paris, 1789) included all of Asia, although the mapping of Tibet was very vague.

Coastal regions were obviously not always well mapped. In Robert's map of the *Archipel des Indes Orientales* (1750) a caption 'Le fond de ce

Golphe n'est pas bien connu' appears for the coastline of the Teluk Tomini in the Celebes (Sulawesi). The *Carte plate qui comprend l'Isle de Ceylon* (1775) includes the captions 'Isles Laquedives dont le détail n'est pas éxactement connu' and 'on ne connoit, ni le nombre, ni la grandeur, ni la situation respéctive des Isles Maldives ... ce qu'on en à tracé ici conformement à quelques cartes manuscrits, ne mérite aucune confiance des navigateurs'. The Australian coast was not fully charted until the Flinders and Baudin expeditions of the 1800s.

The course of transoceanic disputes between European powers was often related to tension in Europe. Colonial borders were usually of secondary importance to the European powers, becoming significant issues only because of other interests and disputes. France and Portugal reached agreement in their 1697–1700 dispute over the area of Maranhao – Brazil north of the Amazon – only after the problem of the Spanish succession came to the fore, and it is possible that French territorial claims for an Amazon frontier for their colony at Cayenne were designed to make Portugal more pliable over the issue. In 1749 the French foreign minister hinted to the British envoy that 'marine' – that is, colonial – disputes could be determined amicably and in Britain's favour in return for her co-operation in European affairs (Shirley 1984; Griffin and McCaskill 1986: 6; Cole Harris 1987: 96–7, 150–1, 168–9; Szarka 1975: 125; Yorke 1749a: 34; Pelletiev 1984: 23–30; J.D. Black 1970–5; Penfold 1974).

The Anglo-Spanish treaty of 1670 had limited Spanish claims to hegemony, permitting the ruler of Britain 'all the lands, countries, etc. he is now possessed of in America'. The foundation in 1732 of a new colony in Georgia, between the Carolinas and the Spanish colony of Florida, led to disputes over whether the new colony was thus legally British (Jenkinson 1785: 197; Lanning 1936). Yet conflict came over competing commercial pretensions in the Caribbean, not over the contested Georgia frontier. Conversely, conflict was avoided in the Anglo-Spanish Nootka Sound crisis of 1790, a dispute arising from competing claims over the western coastline of North America, and specifically the Spanish seizure of British warships trading on Vancouver Island, because the weaker power, Spain, was intimidated successfully. Both Britain and Spain had been concerned to chart the Pacific coast in order to establish their claims more clearly. In late 1662 when an English force arrived at Bombay to take over from the Portuguese, it was unclear whether it was simply Bombay island that they had been ceded, as the Viceroy contended, or, as the English insisted, the entire archipelago. When in 1665 the English force finally negotiated landing rights at Bombay, making heavy concessions in terms of Bombay's territorial extent, these were repudiated by Charles II. Political alliances between Britain and Portugal and Britain and the United Provinces helped to blunt the edge of their colonial disputes, in the Dutch case for the century after the Treaty of Westminster of 1674. British plans to send warships on a voyage of discovery to the 'South Seas' in 1749, a period of hope for better Anglo-Spanish relations, led to Spanish complaints about British interest in the Falkland Islands, which the Spaniards declared they had already discovered and settled:

> he [Spanish minister Carvajal] hoped we would consider what air it would have in the world to see us planted directly against the mouth of the Straits of Magellan, ready upon all occasions to enter into the South Seas, where the next step would be to endeavour to discover and settle some other islands, in order to remedy the inconveniency of being obliged to make so long a voyage as that to China, to refit our naval force upon any disappointment we might meet with in our future attacks upon the Spanish coasts ... I [British envoy Keene] told him it would be difficult to take any step for the improvement of navigation, and procuring a more perfect knowledge of the world in general, that might not be subject to twisted interpretations, and imaginary inconveniencies.

Spain was trying to impose an oceanic frontier to keep the Pacific closed to outsiders and mysterious. The British ministry could not accept this principle, but, as they were seeking improved relations in order to counteract the francophile tendencies in Spanish policy, they decided not to send ships into the Pacific (Cook 1973; Webb 1975; J.M. Black 1994; Keay 1991: 132–3; Holdernesse 1749: 457; Newcastle 1749: 550–1; J.M. Black 1985: 104–5; Bedford 1749; Keene 1749: 177–8, 271, 265–8, 330).

Maps had been employed in Anglo-French differences in the 1680s over the frontier between Canada and the territories of the Hudson's Bay Company (J.D. Black 1970–5: 49–55; Barber 1990: 19), and differing maps played a role in the failure to settle Anglo-French disputes in North America in 1755. This was related to the fact that the area in dispute was inland, and thus poorly mapped. Attempts after the war of 1744–8 to settle the frontier were inhibited by a bellicose and somewhat naïve British public opinion, as well as by the competing views of the two powers. Criticizing ministerial pusillanimity, the *Westminster Journal* of 18 March 1749 argued that 'It behoves us to perambulate exactly the boundaries betwixt us and the French in North America; to determine precisely what is ours, and what is theirs, upon the footing of the Treaty of Utrecht'. The geographical knowledge of the British commissioners appointed to negotiate frontier disputes with France was limited. Joseph Yorke wrote of his colleague William Mildmay,

> As to Mr. Mildmay, I know him very well, and for ought I see he may do very well for one of the commissaries, though it would sure be more decent to nominate somebody, that is more knowing in the geography of America and the West Indies, than I or my supposed colleague.

Small-scale conflict broke out in 1754, as a result of competing claims in the Ohio river valley, and in 1755 both powers simultaneously negotiated and armed for conflict (Yorke 1749b: 103; Pease 1936; Savelle 1940; *Westminster Journal*, 25 March 1749). In January 1755 Mirepoix, the French envoy in London, who had been sent back to his embassy in order to maintain the peace, drew attention to the differences between British and French maps of North America (Mirepoix 1755: 18). Two months later he reported being told by Sir Thomas Robinson, the British Secretary of State,

> que par l'irrégularité du local, la différence de leurs cartes et des notres, et l'infidelité des unes et des autres, il étoit impossible de fixer des lignes justes qui puissent satisfaire aux objets des deux nations et que c'étoit sur cette considération que sa cour proposoit de convenir sur les degré de latitude, la methode la plus seure pour se décider sur des parties aussy peu connues
> (Mirepoix 1755: 261)

Nevertheless, the differing maps were not discarded. In February 1755 the French envoy in The Hague discussed the dispute with Fagel, in effect the Dutch foreign minister,

> La carte de Danville étoit sur ma table. Il l'examina, et aprez un moment de silence, il me dit, 'si les positions sont exactes sur cette carte, il n'y a point de doute que les prétentions de la France, ne soyent légitimes; mais les anglais peuvent avoir des cartes sur lesquelles il serait peut être aussy aisé d'adjuger la raison de leur côte. Il m'en arrive une d'Angleterre'.

Five months later, the French envoy in Hanover discussed British pretensions in North America with the Earl of Holdernesse, Robinson's

Secretarial colleague, with reference to a map of America Holdernesse had (Bonnac 1755: 106–7; Bussy 1755: 22; Clayton 1981; Reese 1988: 274–310; J.M. Black 1990a: 67–77). It is scarcely surprising that the crisis and subsequent conflict created a public market for more maps (Reitan 1985: 54–62). *A Universal Geographical Dictionary; or, Grand Gazetteer* (London, 1759) was, the title-page proclaimed, 'Illustrated by A general Map of the World, particular ones of the different Quarters, and of the Seat of War in Germany'. New maps of America were announced in the issues of the *Daily Advertiser* of 3 August, 5 September and 10 September 1755, while, at the end of the war, the *Universal Magazine* of March 1763 provided a map of the extent of territory Britain now controlled in North America. At the end of the war there was also pressure for a clear-cut territorial settlement. 'Nestor Ironside' stated, in the *London Evening Post* of 23 September 1762,

> let our negotiators take great care, that the bounds of our dominions in all parts of the world, with which the new treaty, whenever it is made, shall meddle, be plainly and fully pointed at; and sure it would not be amiss, if authentic charts or maps were thereunto annexed, with the boundaries fairly depicted. The late peace of Aix la Chapelle [1748] proved indefinite for want of this precaution.

This was indeed the case. The difficulty of settling contested territorial claims had led to their postponement in 1748 when peace had last been negotiated between Britain and France. The principal British plenipotentiary, the Fourth Earl of Sandwich, had been brushed off when he pressed his French counterpart on outstanding disputes:

> As to the Isles of St. Lucia and St. Vincent, and the security to be given for the quiet possession of Nova Scotia and Annapolis, Monseiur St. Severin says nothing will be disputed that is ascertained in former treaties, and consequently that in the general restitutions to be made, those islands are to be evacuated of course, and in the renewal of the respective treatys all possessions which were ceded to us are confirmed to all intents and purposes, but that it can not be adviseable to clog the Preliminarys with too much detail for that if we are to ascertain our right to our possessions in particular, they must do the same with regard to their's, which will lead us into a discussion too tedious for our present circumstances.

> (Sandwich 1748: 486)

The 1763 settlement was much more clear in North America and the West Indies (although there had been problems over the Mississippi boundary in the negotiations), but then the outcome of war had been decisive. A map was joined to the instructions to the British negotiator, the Duke of Bedford, in order to help him negotiate the Mississippi boundary (Rashed 1951: 166 and map opposite 254). After the American colonies became independent, they were involved in frontier disputes with Britain and Spain. Due partly to deficiencies in mapping, the American–Canadian border was drawn in a contradictory manner, leading to serious disputes that were only finally settled in 1842 (Ritcheson 1969; Wright 1975; Stuart 1988). A settlement in 1795 of the disputed frontier with Spanish West Florida brought America much of the future states of Mississippi and Alabama (Whitaker 1927; Bemis 1960). There were also serious disputes over frontiers between individual states. Rivers were very important in the creation of provincial, state and county boundaries in the United States, but they could not be used everywhere. Despite tentative agreements between Pennsylvania and Maryland reached in 1732 and 1739, neither resulted in a permanent solution. In 1763 David Rittenhouse made the first survey of the Delaware Curve, which would not be defined satisfactorily until 1892, and the remainder of the Pennsylvania–Maryland boundary was settled by Charles Mason and Jeremiah Dixon in 1764–7. Violent disputes between Connecticut and Pennsylvania beginning in 1769 were only settled by Congress's acceptance of the Pennsylvanian claim in 1782. The Pennsylvania General Assembly ordered the survey of the northern line in 1785. There were also violent disputes, from pre-revolutionary days, between New York and New Hampshire, over what was to become of Vermont, and they only settled these in 1790. Aside from competing claims to the Western Territory on the part of seven of the founding states, there was also a dispute between Massachusetts and New York (Downes 1970; Darby and Fullard 1970: 200; *The Atlas of Pennsylvania* 1989: 81).

DEFINING FRONTIERS

The problem of defining frontiers played a central role in the causation of conflict in early-modern Europe. Poorly defined boundaries on the ground, and thus territorial divisions that were difficult to represent clearly on maps, were an integral feature of *ancien régime* international relations, themselves a consequence of the 'mind-set' of the period, with its approach to territory in legal/feudal rather than spatial terms, and the related limitations of contemporary mapping. The societies of the period lived with a pronounced degree of tension over frontier zones, areas of overlapping jurisdictions and divided sovereignty. This undifferentiated territorial situation both provoked more serious clashes and became the occasion for disputes arising from other causes. The situation was not static. Maps came to be increasingly used in diplomatic business, there were problems in using the maps because of the way frontiers were depicted on maps, and the portrayal of frontiers on maps changed in response to the new demand. A firmer grasp of the nature of a linear frontier developed, one that was possibly associated with improved mapping and a more definite perception of the nature of political sovereignty, although the notion of such frontiers long predated the improvements in mappings in the seventeenth and eighteenth centuries. For example, Charlemagne's division of his dominions among his three sons in 806 drew on a number of territorial criteria including the linear. The second clause read,

> To our beloved son Pippin: Italy, which is also called Langobardia; and Bavaria as Tassilo held it, except for the two *villae* called Ingolstadt and Lauterhofen which we once bestowed in benefice on Tassilo and which belong to the district called the Nordgau; and that part of Alemannia which lies on the southern bank of the river Danube and the boundary of which runs from the source of the Danube to where the districts of the Klettgau and the Hegau meet on the river Rhine at the place called Enge [near Schaffhausen] and thence along the river Rhine, upstream, to the Alps – whatever lies within these bounds and extends southwards or eastwards, together with the duchy of Chur and the district of the Thurgau.
>
> (King 1987: 252)

In the eighteenth century, in particular, advances were made in mapping and in devising a more spatially territorial approach to frontiers, although this process remained incomplete at the time of the French Revolution. Rivers were used to delimit frontiers in the Peace of Nijmegen in 1678 and the policy was maintained at that of Ryswick in 1697. A stronger interest in precision inspired advances in mapping, which in turn gave the spatial aspects and pretensions of diplomacy a new cartographic precision, though the general problems of cartography remained – the scale of the line on the map, delineation, emphasis through colour and style.

Improved mapping took three forms. First, cadastral maps – maps made for taxation or other administrative purposes. In much of Europe all early large-scale maps were cadastral. They often involved the mapping of estates. This led to the borders of estates being put on maps with great accuracy. Cadastral mapping was employed extensively by the Swedes, both in Sweden and in their German conquests in the seventeenth century. Such mapping was seen as a necessary complement to land registers and thus as the basis of reformed land taxes. The Swedish Pomeranian Survey Commission of 1692–1709 was designed to provide the basis for a new tax system. Detailed land surveys of Piedmont and Savoy, establishing the ownership and value of land, were completed in 1711 and 1738 respectively, while cadastral mapping of Lombardy was carried out in the late 1710s with the backing of the Emperor Charles VI (Helmfrid 1990: 39–42; Quazza 1957; Bruchet 1896; Klang 1977; Andrews 1985; Mason 1990; Haslam 1991: 55–72; Bendall 1992; Thomas 1992; Harley 1992: 141–57). Such mapping helped to lead to more accurate maps of frontiers. Conversely, where such cadastral mapping was generally absent, the detailed mapping of frontiers was delayed. Thus, it was not until the late 1740s that the first detailed delineation of the entire Anglo-Scottish border was carried out. No large-scale cadastral survey was carried out in Britain until the Tithe Commutation Act of 1836 when payments in kind were replaced by monetary rents.

The second major development was the growing importance of large-scale military surveys. The Austrians, who ruled Sicily between 1720 and 1735, used army engineers to prepare the first detailed map of the island. The French military engineers of the period, such as Pierre Bourcet, tackled the problems of mapping mountains better than their Piedmontese counterparts, creating a clearer idea of what the alpine region looked like. Following the suppression of the '45, Lieutenant-Colonel David Watson, Deputy Quartermaster-General to the forces, assisted by William Roy, prepared between 1747 and 1755 the map known as the 'Duke of Cumberland's map' of the mainland of Scotland, which was based on a military survey of Scotland (Moore 1988: 28–44; O'Donaghue 1977; d'Albissin 1979; Solon 1982; Buisseret 1982; Pallière 1979: 51–66; Baillou 1984: 116; Buisseret 1985: 72–80; Pallière 1985a: 59–67; Pallière 1985b: 39–45; Pallière 1986: 50–67). A major military survey of Bohemia was begun in the 1760s and completed under Joseph II. Lower Austria was surveyed from 1773 and an enormous survey of Hungary completed in 1786 (Dickson 1991: 617). Frederick II had Silesia mapped.

The third major development was the improved measurement of

longitude. Until the eighteenth century there were no clocks accurate enough to give a ship's meridional position and longitudinal mapping faced problems. Many islands were placed too far to the west or the east, and this caused shipwrecks, for example on the Scilly Isles. Coronelli's map *Route maritime de Brest à Siam et de Siam à Brest* was based on the Jesuit mission sent to Siam (Thailand) by Louis XIV in 1685. It carried a note saying that the map employed two sorts of longitudinal markings, those generally agreed and those based on information from the Jesuits. A French map of 1739, *Carte de l'Ocean Oriental ou Mers des Indes dressée au dépost des cartes plans et journaux de la Marine comparée avec la carte Hollandoise de Ptetergoos et la carte Angloise de Thornton,* revealed major differences.[3]

In response to an Act of Parliament of 1714 offering a reward for the discovery of a method of determining longitude at sea, John Harrison devised a chronometer that erred by only 18 miles on a return journey to Jamaica in 1761–2. Progress on land was swifter. In 1679–83 the French Académie had worked out the longitudinal position in France. A geodetic survey of France was carried out. In 1708–18 the Jesuit Fr Regis supervised the first maps of the Chinese Empire to be based on astronomical observation and triangulation. An improved ability to measure longitude affected mapping, obliging and permitting the drawing of new maps (Gallois 1909: 193–204; Brown 1941; Guyot 1955; Forbes 1974; Howse 1980; Howse 1990: 86–114, 160–83). The *Atlas Geographus* published in London in 1740 noted that

> the curious, by casting their eye on the English map of France, lately done and corrected according to the observations of the Royal Academy of Sciences at Paris, may see how much too far Sanson has extended their coasts in the Mediterranean, the Bay of Biscay, and the British Channel.
>
> (*Atlas Geographus* 1740: 979)

In the 1760s the border between Portuguese Brazil and Spanish South America was surveyed and mapped in detail, because it was realized that the old maps were wrong.

A map formed part of an Austro-Dutch treaty in 1718 that delineated the frontier between the United Provinces and the Austrian Netherlands. This owed much to the publication in 1711 of the Fricx map of the Low Countries, the first relatively large-scale military map of Europe, and the predecessor of the work of Cassini for France and Roy for Scotland. The frontier was fixed literally on a map signed and sealed by plenipotentiaries as an annex to that treaty. This practice became established by the end of the century. The French foreign office created a geographic section in 1772 and in 1780 it acquired the collection of about 10,000 maps of the famous geographer d'Anville (Konvitz 1987; Sahlins 1989; J.M. Black 1990b: 192–7; Sahlins 1990: 1423–51; Watelet 1990/4: 33–49).

The habit of referring to maps increased. Maps had been used in the recording of European frontiers since at least the fifteenth century, and had of course been used in antiquity, with the Egyptian map of Nubian gold mines, Ptolemaic maps and Roman cadastral maps. One map was drawn to show a small section of the frontier between France and Burgundy in 1460; another to show the frontiers of the kingdom of Naples in the late fifteenth or early sixteenth century (de Dainville 1970: 99–121, for example 112 and fig. 10; Almagia 1929: 13, pl. xiii; Harley and Woodward 1987). Maps were increasingly used from the sixteenth century. A map was used for the negotiations that led to the Anglo-French Treaty of Ardres of 1546. These negotiations were not without serious difficulties. Before the treaty William, Lord Paget, and a French emissary went with several guides to examine the source of a proposed boundary stream and fell into a serious dispute over which of five springs was the source of the river (Shelby 1967: 94–101; Barber 1991: 145–51; Gammon 1973: 106; Tyacke 1983; Harley 1987; Harley 1988: 57–76; Helmfrid 1990: 34–9; Buisseret 1992). In 1712, during the successful negotiations over ending the War of the Spanish Succession, Torcy, the French foreign minister, urged his British counterpart to look at a map in order to see the strategic threat posed by the alpine demands of Victor Amadeus II of Savoy-Piedmont:

> Ce seroit ouvrir le Royaume, en donner les clefs à M. le Duc de Savoye et laisser le Dauphiné à sa disposition que de luy abandonner ce qu'il demande au delâ du Rhone, Briançon, et le fort Barraux. Prenez la peine, Monsieur d'examiner seulement la carte du pays.

In 1750 Puysieulx, a successor of Torcy, informed a French envoy that 'Il ne faut que la simple inspection d'une carte géografique pour connoitre qu'il est de l'intérêt de l'Electr. Palatin de se tenir toujours uni au Roy'. One measure of the growing importance of maps was sensitivity about allowing copies to be made (Torcy 1712: 358; Puysieulx 1750: 282; Chambers and Pullan 1992: 405–6; *Daily Universal Register*, 11 August 1786).

Maps were increasingly referred to in crises and wartime by diplomats and politicians. In 1718 Reeve Williams published his defence of

31

British foreign policy, *A Letter from a Merchant to a Member of Parliament, Relating to the danger Great Britain is in of losing her trade, by the great increase of the naval power of Spain with a chart of the Mediterranean sea Annex'd*: 9,000 copies of the pamphlet were printed. The inclusion of the map helped to make the work interesting. The *Worcester Post-Man* reported that

> a notable book was delivered to the Members of Parliament, with a chart annex'd of the Mediterranean Sea, whereby it demonstrately appears of what importance it is to the trade of Great-Britain; that Sicily and Sardinia shall be in the hands of a faithful ally, and if possible not one formidable by sea. That these two islands lie like two nets spread to intercept not only the Italian but Turkey and Levant trade.

Sir James Harris, British envoy at The Hague, recorded of the Cabinet meeting he attended in London on 23 May 1787, as the Dutch crisis of 1787 neared its height, that the Master General of the Ordnance, the Third Duke of Richmond, 'talked on military operations – called for a map of Germany – traced the marches from Cassel and Hanover, to Holland, and also from *Givet to Maestricht* [*sic*]'. Thus the possibilities of French and British-subsidized German intervention were outlined to the Cabinet by the use of a map. The following day Harris saw William Pitt the Younger. He recorded that Pitt 'sent for a map of Holland; made me show him the situation of the Provinces etc.' George III used a map to follow the Prussian invasion of France in 1792 (J.M. Black 1986: 53–8; Barber 1990: 1–28; Malmesbury 1844: 304, 306; Grenville 1792: 37–9; George III 1792: 37–9; Finch 1741; Poyntz 1746).

In 1758 Lord George Sackville had written to Holderness about military operations in Germany, 'You will see Cappenburgh in the map' (Sackville 1758).

Maps were consulted more often in the eighteenth century in relation to political discussions and in diplomatic negotiations, and were on the whole geographically more accurate. This did not, however, make negotiations easier unless the two powers were prepared to compromise. Accurate maps did not help if negotiators disagreed over the terms that described frontiers in treaties. Such maps did not necessarily help the definition of borders in detail. Improved mapping could help to reveal differences of opinion. On the other hand, if rulers were seeking compromise, improved mapping could help to cement agreement, and such an attitude of conciliation and rational arrangement characterized French policy between 1748 and 1789. In addition, if it is argued that conflict arose as much from accident and from the opportunism latent in an international system that was imprecise in a number of spheres – most clearly frontiers – as from deliberate planning for war, then increased precision in the mapping of frontiers was as important as the related consolidation of territorial sovereignty and increasing state monopolization of organized violence. All were different facets of the consolidation and spread of governmental authority and the erosion of the distinctive features of border zones. The implementation of firm frontiers was bound up with the existence of more assertive states and growing state bureaucracies, which sought to know where exactly they could impose their demands for resources and where they needed to create their first line of defence. As a result, the period from the late middle ages witnessed a burgeoning emphasis on frontiers throughout Western Europe, despite the constraints produced by the limited extent of contemporary mapping. Changes in warfare also affected boundaries and maps. Fighting between ships of the line on open seas did not generate large-scale mapping, but, rather, improvements in instrumentation, with the sextant and the chronometer. On land, fortifications and garrisons provided an opportunity for large-scale mapping of border regions and mountain passes.

Improved mapping helped to make the understanding of frontiers in linear terms, rather than as zones, easier and thus played a major role in frontier negotiations, not least in the attempt to produce more 'rational' frontiers by removing enclaves. This was a hesitant process. Henri Arnault de Zwolle, councillor of Philip the Good of Burgundy, produced in 1444 a map of a contested region between France and Burgundy. This was seen by Duke Philip as part of a process by which French enclaves could be defined and eliminated in order to simplify the frontier. Frontier rationalization, however, was a hesitant process in Western Europe. The notion that Louis XIV wished to create a rationalized frontier between France and the Spanish Netherlands has been queried (Richard 1948: 112; Ekberg 1979: 118–19; Sahlins 1990: 1433–4), and, instead, it appears that he wished to use feudal claims to push forward the zone of French power. Far from seeking the stability of a clear line, Louis wanted to leave the frontier amorphous and his pretensions ambiguous, so that his power could be advanced as opportunities offered. Even if the impact of Louis XIV's activities was a clarification of the frontier, his expansionism was opportunistic. Enclaves were retained where it suited French interests. Mülhausen remained as an ally of the Swiss Confederation, an enclave in French Alsace, until the 1790s, because the Bourbons did not wish to upset

33

their relations with the Swiss. French fortresses were established to the east of the Rhine and Alps, so that any notion of the role of fortifications in helping to define frontiers has to be qualified by a realization that they could also lend force to enclaves, and thus sustain the zonelike nature of some frontiers. Louis XV wished to conserve French enclaves, for example the towns of Beaumont and Gimay in the Austrian Netherlands. Puysieulx complained in 1748 that the cession in Louis XIV's later years of Pinerolo, Fenestrelle and Exilles, all east of the Alpine watershed, closed the doors of Italy to France. The lengthy negotiations between France and the Prince-Bishops of Liège over Bouillon between 1697 and the mid-eighteenth century were scarcely characterized by a spirit of 'reason' insofar as the latter term is now understood

> Dans les multiples 'factums' qui vont s'échanger jusqu'au milieu du XVIIIᵉ siècle, les deux parties traiteront en long et en large cette question de Bouillon, en remontant souvent jusqu'au XI siècle. Elles invoqueront en général quantité d'autorités que nous jugerions aujourd'hui fort sujettes à caution: récits plus ou moins légendaires ou merveilleux du haut Moyen Age, témoignages d'auteurs anciens qui n'écrivent que sur des oui-dire, généalogies plutôt fantaisistes, etc.

(Louis XV 1745: 77; Puysieulx 1748; Harsin 1927: 161–73, 164; Sack 1986; Bonenfant 1953: 73–9; d'Albissin 1966: 390–407; Allies 1980)

On the other hand, more 'rational' methods of establishing disputed frontiers did not necessarily ease frontier tensions. Improved mapping could highlight differences and thus, for example, exacerbate problems with enclaves if one party wished to create or sustain difficulties. Alongside change, there was a persistence of traditional attitudes. Despite the general conception of sovereignty in jurisdictional, rather than territorial, terms in seventeenth-century Europe, the idea of natural frontiers (readily grasped geographical entities), principally mountains, had become 'a widespread dictum of geographical discourse', readily employed in diplomatic discussion. In the 1659–60 negotiations over their new frontier in the eastern Pyrenees, both French and Spanish negotiators advanced geographical claims as well as historical arguments, though the eventual frontiers 'were neither historical nor geographical but rather a compromise resulting from a bitter diplomatic struggle'. The frontier continued to give rise to negotiations and disputes. In 1688 the Spaniards proposed that a section of the border at

Aldudes in the western Pyrenees, currently undivided and common to both nations, should be partitioned. Negotiations over the Pyrenean frontier were, however, still encountering serious difficulties in 1775 and were not settled until 1785, and it was the French Revolution that led to a new departure, with the 'politicization of natural boundaries' and the instilling of a stronger and more firmly policed sense of national territoriality in border regions (Sahlins 1989: 35, 49, 187; Rebenac 1688: 56; Grantham 1775: 41, 50–1).

The selection of natural frontiers was not, of course, free from serious problems. These affected both rivers and mountains. Rivers were not on the whole canalized, as many were to become in the nineteenth century. Their courses shifted, islands were created and disappeared and river courses could be affected by drainage works. Bartolo de Sassoferrata, in his treatise *De Fluminibus seu Tiberiadis* (de Dainville 1970), had considered the problems of meanders, changes of river course and new islands in rivers. In his treatise he included cartographic solutions to the changes of boundaries caused by alluvial processes. His thesis was very influential in French judicial geography. After the Act of Union of 1707 the shifting channel of the Solway was accepted as part of the Anglo-Scottish frontier. In 1719 works on the Elbe led to a serious dispute between Hanover and Prussia. Disputes between the Duchies of Parma and Milan were recurrent, being raised at the Congress of Cambrai in 1723 and becoming a problem again in 1733, and 1789–90, when the principal Bourbon powers, France and Spain, exerted pressure on behalf of Parma (de Dainville 1970: 118; Hetherington 1991; Whitworth and Polwarth 1723: 150; Robinson 1733; Noailles 1789; Noailles and Llano 1790; Kaunitz 1789; Noailles 1790: 357, 44, 358, 51, 71–2, 285–96, 359, 202). In 1762 an island in the Ticino led to a border dispute between Milan and Savoy-Piedmont. Rivers were, however, extensively used as the basis for frontiers, both old and new. At the end of the Russo-Swedish war of 1741–3 Russia by the Treaty of Åbo secured a triangular slice of south-east Finland, based on a new river line. The Swedo-Finnish boundary at the head of the Bothnian Gulf remained on the line of the Kemijoki until 1809 when the Russians pushed it west to the Tornionioki. River lines, including the Dvina, Niemen, Bug and Vistula, were used in the three partitions of Poland in 1772, 1793 and 1795.

Defining mountainous frontiers could be difficult, not least because it was unclear how they should be defined. How to show relief in medium-scale mapping has always been a problem – one that has never really been solved. Disputes were a major theme in Franco-Savoyard negotiations after the Peace of Utrecht (1713). The convention of Paris of 4 April 1718 left several issues outstanding. Negotiations in the late 1750s, culminating in the Treaty of Turin of 24 March 1760, marked a step forward. Eight maps, defining the watershed, played an integral role in the treaty, having the same weight as the text of the treaty (Manno, Vayra and Ferrero 1886–91; Pallière 1986). Nevertheless, until the mid-nineteenth century it was very difficult to map mountainous areas. Serious problems were encountered in getting people on to the mountains and there were also difficulties with topographic mapping.

Elsewhere, moves towards more defined frontiers can be discerned. This took two major forms, the move towards undivided sovereignty and that towards neat linear boundaries. The former process was handicapped across a broad swathe of Europe by the constitution of the Holy Roman Empire and by such historical legacies as shared authority, for example between the United Provinces and the Prince-Bishopric of Liège in Maastricht, or the post-Westphalia alternative inheritance of the prince-bishopric of Osnabrück between an elected Catholic prelate and a member of the Protestant house of Brunswick. In the Empire, single maps were not a good way of showing princely territorial rights. It was usually beyond the ingenuity of even the most skilful cartographer to indicate on one map areas of mixed jurisdictions, owing allegiance to different rulers for aspects of their existence (for example, Schleswig-Holstein). Nevertheless, the mapping of the Empire improved. George III commissioned a major survey of Hanover in the 1760s and Holstein was mapped anew towards the end of the century. A survey of Holstein, Lübeck and Hamburg was published in 68 sheets between 1789 and 1806.

FRENCH INNOVATIONS

Considerable success in making sovereign powers more consistent was achieved. In Alsace the ambiguous relationship established under the Peace of Westphalia between the French crown and the ten Alsatian towns known as the Decapolis was defined in accordance with Louis XIV's power. The principality of Orange was also acquired by Louis.

The move towards frontier lines rather than zones was especially marked in the case of France, not least because it contrasted so obviously with the complex overlapping of jurisdictional authorities that had arisen because of the impact of French power in the western borderlands of the Empire. This had served French interests as a means of providing opportunities for territorial expansion, most obviously with Louis XIV and the *réunions*, but in the eighteenth century *ancien régime* France ceased to be so interested in European territorial expansion, and this new attitude was linked to a desire for the stabilization and, thus, in part, rationalization of frontiers. In 1750 Louis XV ordered Marshal Belle-Isle to settle amicably all the disputes over the frontier between newly acquired Lorraine and the Empire. Some twenty-five years later the ministry of foreign affairs gained jurisdiction over boundary matters from the ministry of war, and established a topographical bureau for the demarcation of limits. The context of this stabilization was, nevertheless, one of French dominance over their neighbours who, especially in the first half of the century, complained about the way in which the French used their strength to secure favourable frontier rectifications (Sahlins 1990: 1438; Barber and Black [unpublished]; Hansen; Whitworth 1715; Sahlen 1721: 56; Ludwig 1726: 61; Christian IV). Nevertheless, claims that the French were seeking drastically expanded frontiers were inaccurate. The *Newcastle Courant* of 20 December 1746 might claim that

> Their ambition is immeasurable, never at rest, making war upon war without reason, justice, or end, to round their dominions, as they impudently term it; and to recover their pretended ancient, but imaginary boundary; all the countries between them and the Rhine one way, and between them and the North Sea, the other.

However, France's principal continental gain in the pre-revolutionary eighteenth century, the Duchy of Lorraine, was a response to the dynastic union of the Lorraine family and the Habsburgs, and can be seen as a defensive step (J.M. Black 1988: 365–70; Ranum 1991: 334–7). In addition, the 'Diplomatic Revolution' of 1756, the Franco-Austrian alliance that lasted, albeit with varying intensity, until 1792, ushered in a period in which French sensitivity over her frontiers diminished and it became possible to negotiate satisfactory solutions to a number of problems. During the unsuccessful Anglo-Dutch attempt to negotiate a settlement of the Austro-French War of the Polish Succession (1733-5) it had been argued that border issues between France and the Empire in Flanders and Alsace would be 'hard to adjust' (Conference of 25 December 1734: 56), but must be part of a general pacification. This had not actually been the case: peace treaties generally had to be negotiated too speedily to provide for such agreements, other than by leaving decisions to commissioners who commonly found it impossible to settle matters. The legacy of border disputes on and near France's eastern frontier was extensive, and in the 1720s and 1730s included a Franco-Palatine dispute over the Alsace frontier and a Franco-Austrian one over that near Saint-Hubert (Archives du Ministère des Affaires Etrangères, Mémoires et Documents; Morville 1725; Chavigny 1727; de Silly 1727; Chauvelin 1728).

Tensions were progressively eased, however, and the Franco-Austrian alliance can be seen as central to the general pacification of Western Europe. In 1749 frontier differences with Geneva were settled by a treaty with subsequent delimitation agreements in 1752 and 1763, which was followed in 1754 by the Treaty of Turin between Geneva and Savoy-Piedmont, one that had taken since 1738 to negotiate (Waeber 1974: 42–3). The French reached settlements with the Austrian Netherlands (1738, 1769, 1779), the Prince of Salm (1751), the Duke of Württemberg (1752, 1786), Prussia over Neuchâtel (1765), the Bishop of Liège (1767, 1772, 1773, 1778), the Prince of Nassau-Saarbrücken (1760), the Duke of Zweibrücken (1766, 1783, 1786), the Canton of Berne (1774), the Elector of Trier (1778), the Prince of Nassau-Weilburg (1776) and the Bishop of Basle (1779, 1785) (Praslin 1763: 54; Murphy 1982: 454; Sahlins 1990: 1438–41; de Lapradelle 1928: 45 n.1; Noel 1946: 336–7; Livet 1966: 58, 228–46, 254–6, 278–93; Nordman and Revel 1989: 29–169). Although older usages persisted, the period also saw the beginnings of a different attitude to frontiers, one that definitely predated the French Revolution. If the right of the strongest still played a major role in the fixing of frontiers, an entirely different principle appeared, that of strict equality between the parties, whatever was the degree of their relative power, both in the course of the negotiations and in the final agreement. The application of this principle was considerably helped by the existence of natural obstacles (rivers and mountains). Using these to draw frontiers offered the possibility of establishing equality in negotiation between the parties in order to create what were termed 'natural' frontiers.

This was certainly the case with the Sardinian frontiers, and can be seen in the negotiations leading to the Franco-Sardinian convention of 1718, the Treaty of Turin between the two powers in 1760, and the delimitation in the Treaty of Worms of 1743 of a new frontier along the middle of the principal channel of the river Tessin between Lombardy and Piedmont. In order to cope with the problems of different channels, the Austro-Sardinian convention of 1755 and Articles 3 and 9 of the Treaty of Turin both stipulated that the principal channel should be followed and that it should be divided in the middle. Charles Emmanuel III of Sardinia (1730–73) was not prepared to rely simply on the text of treaties, which were never completely precise. He wanted them represented materially, by a drawing – in other words, a map – and, on the ground, by a continuous demarcation of boundary marks and posts. In 1738 he created the Ufficio degli Ingenieri Topografi. The new Franco-Sardinian frontier was not only 'natural' but also linear. The Convention of Turin of 1760 fixed the frontier with Monaco, and that of 1766 with Parma, while the frontier with Genoa was clarified in the Seborga, and between 1770 and 1773 Antoine Durieu and the Genoese Gustavo Gerolamo produced joint maps of the hitherto-contested frontier. The frontier with Switzerland in the area of the Great Saint-Bernard was not, however, settled until 1940. The cartographic work of Charles Emmanuel was very extensive: he began with the mapping of the Duchy of Savoy in 1737 and finished with that of his entire territories in 1772.

The same process was at work elsewhere. Better maps were generally not available for use in international relations until the mid-eighteenth century. The demand–supply question is a subtle one: diplomats also sponsored mapping enterprises, political purposes and values shaping the content and uses of maps so that a transition from a juridical to a cartographic depiction of boundaries – the former according to lists, the latter according to lines – could take place. In 1743 the Austrians demanded both to see old maps and that the Republic of Venice name an experienced mathematician to work in concert with their own in order to settle the frontier (Montaigu 1743: 25). Nine years later the British envoy in Vienna recorded:

I saw Monsieur Ulfeld [Austrian Chancellor] likewise yesterday, and found him with several papers, lying before him, which, he said, were three different treaties with the Venetians, and by which a final end was put to all the disputes about frontiers etc. between the House of Austria, and the Republic of Venice, which had, upon many occasions, given so much trouble to both those powers, and some of which had subsisted, for two hundred years. Monsieur Ulfeld attributed a great deal of the merit of this matter to Monsieur Tron, the Venetian Ambassador, who, indeed, has acted a wise and able part, in all the late transactions between this court and the republic. It is certain likewise, that it was a great happiness for the Empress Queen [Maria Theresa], to have such a minister to employ, on her side, in this affair, as Count Christiani, the Great Chancellor of Milan, who, besides his very great abilities and knowledge, carries always along with him that spirit of conciliation, which is so necessary, to bring business of this nature to a conclusion.

(Keith 1752)

There was an important settlement between Sweden and Denmark-Norway in 1751. This laid down the boundary between the two states, ending Swedish-Norwegian 'common districts' in the interior of Finnmark. Maps were used in the protracted and difficult negotiations over the Austro-Turkish frontier that followed the war between the two powers in 1788–91. The two powers eventually concluded a convention that ceded Orsova and a stretch of land on the upper Unna to Austria. The convention, signed on 4 August 1791, referred to lines on a map (Roider 1982: 177, 189; *Gentleman's Magazine* 1791: 861). The following day Sir Robert Murray Keith, the British mediator, reported:

The Imperialists had only three copies of the map of the frontiers of the two Empires (which is so often mentioned in the recent Convention,) these they have given to the Turks, and to the Prussian Minister. But they have engaged to deliver to each of the mediating ministers, on our return to Vienna, a correct map of that kind with all the limits carefully marked out, according to this last adjustment. I shall think it my duty to send that map to your Lordship, as soon as it shall be put into my hands.

(Keith 1791: 167)

Yet there were still serious problems affecting the peaceful settlement of frontier disputes, and they could still be manipulated by aggressive rulers, as the Emperor Joseph II demonstrated over the Dutch–Austrian Netherlands frontier in 1784. A major source of tension, too often minimized in accounts that centre on the initiatives of chancelleries, was the role of local action. Disputes between local communities accentuated problems. A violent territorial struggle between two frontier communities led Victor Amadeus II to stress the need for a settlement of Savoy-Piedmont differences with France in 1716: 'nous avons eu nouvellement connoissance que la communauté de Guillaume en Provence, agit, par voie de fait, contre celle de Châteauneuf dans la vallée d'Entraunes, sur ce que celle-là prétend d'étendre son territoire sur celle-ci'. The king stressed the need

de terminer les différends des confins, en conséquence du traité d'Utrecht, afin d'assurer, de plus en plus, la bonne union entre les cours; et, faisant cesser toutes les contestations entre les sujets, maintenir, par ce moyen, la bonne harmonie à tous égards.

(Instructions 1716: II, 148)

The long-standing conflict between two Apennine communities, Mioglia and Sassello, played a fundamental role in disputes between Savoy-Piedmont, Milan and Genoa in the first half of the century, and the frontier issue was essentially a matter of the dispute at the local level, with only periodic interventions by the relevant central governments (Grendi 1986: 811–45). The role of local attitudes and actions in the eastern Pyrenees emerges clearly from studies of that frontier (Sahlins 1989).

Another source of difficulty was the growing determination of governments to control their frontier regions. This was scarcely new, but the states of the eighteenth century, with their larger and better-controlled armies, were better able to achieve this than their predecessors had been. In 1763 Tillot, the leading minister in the Duchy of Parma, organized a small expedition against Mezzano, an episcopal fief whose population resisted integration with the duchy. In 1767–8 he took analogous police measures over the Corti di Monchio, a mountainous region on the Parmesan–Tuscan frontier, where the privileges and immunities of the local ecclesiastical lord had also created difficulties, the area containing many deserters and smugglers. Such steps were symptomatic of the active pursuit of pretensions that was an obvious feature of the period. In 1768 the Duchy of Parma also pushed her claims to Bozzolo and Sabbioneta. Such claims were not new. Parma, like many other states, had a portfolio of territorial claims, many of considerable antiquity, which she pushed whenever she had the opportunity (J.M. Black 1990c: 353–64; Bédarida 1928: 2; Greengrass 1991). Such actions naturally aroused concern on the part of other rulers, incited or exacerbated disputes and made conciliation more difficult. The opportunistic use of pressure was demonstrated by the successive reinterpretations of Austrian and Prussian gains in the first Partition of Poland. The latter probably benefited from a map that was secretly prepared by Prussian officers (Topolski 1969: 81–110).

A related problem was posed by the diminution in buffer zones. An obvious visual contrast between a modern map of Europe in 1500 and another of Europe in 1700, 1750 or 1789 is that later there are fewer independent polities. When combined with the growing determination of these polities to control their frontier zones, the bellicose nature of society and the opportunistic character of international relations with the particular stress on the value of brinkmanship, it is scarcely surprising that frontier disputes were a source of conflict (J.M. Black 1987). On the other hand, the lessening in the number of independent 'players' and the diminution in buffer zones can also be seen as having

41

the opposite effect. This was also related to the relative shift in the European 'states system' to Eastern and Central European powers, Russia, Austria and, although to a lesser extent, Prussia, a shift that was readily apparent from the 1680s (J.M. Black 1990b). France and Spain had waged their myriad conflicts between 1494 and 1659 in large part through or at the expense of intermediaries, as they struggled for control of the buffer zones of northern Italy and the lands between the Marne-Saône and the Rhine, regions of divided sovereignty, numerous, weak polities and complex and contentious frontiers (J.M. Black 1992: 91–114). In such a struggle, which was essentially continued between France and Austria until 1756, the fate of small territories such as Montferrat or Guastalla played a large part in diplomacy and the essential unit of diplomatic exchange was jurisdictional-territorial not geographical-territorial. This was reflected in the dominance of succession disputes in the international relations of the period. Most of the major wars in Western Europe in the pre-revolutionary eighteenth century were succession conflicts – the Wars of the Spanish, Polish, Austrian and Bavarian Successions. The Seven Years War can be seen as an attempt to reverse the principal territorial consequence of the War of the Austrian Succession, and thus as an extension of it.

EASTERN EUROPE

In Eastern Europe, in contrast, geographical-territorial issues played a larger role. The major states lacked good historic claims to the areas in dispute, the texture of sovereign polities was less dense than in Western Europe – not least because hitherto autonomous regions, such as the Ukraine, were brought under control (Subtelny 1986) and dynastic succession was not the major diplomatic idiom nor generally a means by which large areas of territory changed hands and through which relative power could be assessed. This owed much to the impact of the Turkish advance, while the fact that Poland-Lithuania became a clearly elective monarchy in 1572 was also of consequence. The Habsburgs gained the throne of Hungary by dynastic means, but had to fight the Turks in order to bring their claims to fruition. Succession disputes were not the issue at stake between Russia and Prussia in the mid-eighteenth century. However, there were differences between Eastern and Western Europe before the Turkish advance.[4] The general European trend towards more defined frontiers was responsible for much of the warfare in the fourteenth and fifteenth centuries, since lands whose status had been ill-defined for centuries were claimed and contended for by rival states

(Bartlett and Mackay 1990). This ensures that compilers of modern historical atlases can begin to draw maps with reasonable certainty that they corresponded to political reality at any given point; but the Christian states in the Balkans – Serbia, Bulgaria, even Hungary – were backward in this respect, and their boundaries remained vague, shifting according to the interests of the local authorities.

Although it might seem that Turkish expansion would end this problem, since the Ottoman Empire was a state as sophisticated as those in Western Europe, in practice the problem remains for the fifteenth century, because the Turks used 'gradualist' methods of conquest. Christian territories on the borders of lands fully assimilated into the Empire (in the sense that the provincial system of government was implemented) might be compulsorily allied to the Sultan or endure tributary status to the Turks. The territories coming into the latter category were in practice under the Sultan's control: he could move his armies there at will, and demand resources and manpower on the same level as within the Empire proper (Inalcik 1954: 103–29). This state of affairs is and was very difficult to depict cartographically. For example, the tributary principality of Wallachia constituted a substantial part of the Ottoman presence in Europe. It would, however, be necessary to give a large amount of detail about relations between the Prince and the Sultan to gain a realistic impression of how powerful the Turks were at any given point beyond the Danube.

There were major territorial changes in Eastern Europe from the late seventeenth century. The Turkish empire ceased to be a major threat and became, to a certain extent, an open frontier for Russian expansion. Russia gained the khanate of the Crimea (1783) and pushed her frontier to the Dniester (1792); Poland regained Podolia (1699). Austria gained Hungary and Transylvania (1699, 1718), Venice (temporarily) gained the Peloponnese (1699). In the eighteenth century Poland became another internal frontier, and it was eventually partitioned by Austria, Prussia and Russia (1772–95).

These major redistributions of territory reflected the volatility of European international relations and the impact of military strength. They also suggest that a concentration on frontier problems as a cause and *modus operandi* of international tension, as might be more appropriate in Western Europe, is less so further east. Frontier infringements did play a role there, for example in the outbreak of Russo-Turkish hostilities in 1735 and 1768. Competing Russian and Turkish interests in the Caucasus buffer zone helped to increase tension between the two powers in the period preceding the outbreak of war in 1787

(Lang 1957: 183–211; Atkin 1980: 29–30; *Cambridge History of Iran* 1991: 327–8; Fisher 1970: 154). The idiom of these disputes was, however, geographical-territorial, not jurisdictional-territorial, and they were not central to the aspirations and policies of the competing powers.

NEW CRITERIA

It was to be the French Revolution that recast the territorial world of Western Europe. Hitherto there had been signs of change, not least in improved measurements and a desire to measure. In Saxony, for example, a new standard mile was introduced in 1715. The geometrician Adam Zurner carried out a cartographic survey, completing 900 maps, which formed the basis of the Electoral Postal Map, first published in 1719. Across much of Europe, improved measurements were matched by a desire to survey and measure which led to detailed land surveys (J.M. Black 1990c: 79–82).

The Revolution replaced jurisdictional-territorial criteria when completely redrawing frontiers in Western and Central Europe. For a society that created a new calendar and a new unit of measurement (the metre), and created new administrative boundaries within France itself, such a spatial reinterpretation was not a bold conception, although in many respects the Revolutionary period should be seen as the culmination of new ideas that developed during the Enlightenment (Ozouf-Marignier 1984: 58–69; Ozouf-Marignier 1986: 1193–213; Ozouf-Marignier 1989; Konvitz 1990: 3–16). Thus the Treaty of Basle of 1795 with Spain used the Pyrenean watershed as a frontier. Furthermore, a stress on new frontiers seemed necessary in order to guarantee the new order in France. The decree of 4 August 1789 which abolished the feudal regime in France affected the Alsatian rights of a number of German rulers including the Dukes of Württemberg and Zweibrücken, the Margrave of Baden, the Landgrave of Hesse-Darmstadt, the Electors of Cologne and Trier and the prince-bishops of Basle and Speyer. This played a major role in exacerbating Franco-German relations, as the Emperor Leopold II and the Imperial Diet took their stand on the sanctity of treaties (Montmorin 1790: 187; Ludwig 1898; Blanning 1986: 102).

In rejecting the *ancien régime*, France, because of the presence of hostile neighbouring polities, had (unlike the United States) to ensure a degree of security that required the creation of a new political order in these polities. This order could incorporate former frontiers as *ancien régime* polities took on a new existence – Genoa becoming the Ligurian Republic, Tuscany the kingdom of Etruria – but even when this was the case there were modifications, while elsewhere there were major changes, as any comparison of the map of Central Europe and northern Italy in 1786, 1797, 1803, 1806 and 1810 would reveal (*New Cambridge Modern History XIV. Atlas* 1970: 132–7). New frontiers, such as those of Etruria, were mapped, while Napoleon's invasion of Egypt and Palestine led to the first accurate map of the region, the French army carrying out the surveying. The Napoleonic regime and its wars provided major stimuli for the detailed mapping of Europe, in order to serve political, financial and military purposes, and to satisfy the quest for information that the regime displayed (Perrot 1977; Perrot and Woolf 1984; Bourguet 1988; Woolf 1991: 87–90). The Revolutionary and Napoleonic wars also stimulated mapping by France's enemies (necessarily so as suitable maps were lacking, Hall 1992: 49), not least in Britain by the Board of Ordnance as concern developed about possible French invasion (Ravenhill 1991: 20–1; Oliver 1991: 120; Seymour 1980; Hodson 1987: 21–30).

Possibly as crucial a change was the continuing but accelerated decline in the number of sovereign polities and the related decline in the number that could be considered as independent states. In Western and Central Europe this owed much to the Revolutionary and Napoleonic period, although that also witnessed the creation of new units, such as the republic of Krakow (1815–46). Furthermore, the defeat of Napoleon led to a reversal of the trend which had culminated in 1810–12 with much of Europe, including the Netherlands, Hamburg, Lübeck, Genoa, Tuscany, Savoy-Piedmont, the Papal States, Trieste, Dalmatia and Catalonia being part of France, while client states, such as the new kingdoms of Westphalia, Bavaria and Italy, were similarly engorged. Nevertheless, although there were more states in Western and Central Europe after the treaty settlements of 1814–15 than there had been in 1812, there were still fewer than there had been in 1789.

In Eastern Europe this process had begun earlier. Independent states had been swept aside by the rise of empires, most recently those of Turkey and Moscow. Hitherto independent territories, including those of the Teutonic Order along the eastern Baltic, independent Russian principalities, such as Novgorod and Pskov, and Islamic ones, such as the khanates of Kazan and Astrakhan, and Balkan states such as Bulgaria, Moldavia, Wallachia, Serbia and Bosnia, were all conquered. The absorption of independent or autonomous polities by larger neighbours continued in Eastern Europe in the eighteenth century and in the

Napoleonic period. Poland was partitioned, and the khanate of Crimea was annexed by Russia in 1783, the kingdom of Georgia following in 1801.

Between the Ottoman overrunning of much of the Balkans in the late fourteenth century and the unifications of Germany and Italy in the third quarter of the nineteenth, many European frontiers became, as in Spain after the dynastic union of Castile and Aragon or Britain after the union of Scotland and England, essentially domestic-political, most commonly judicial and financial, rather than of any international significance. This process was especially marked in Western Europe, with its denser and more historical fabric of jurisdictional authorities, and the accompanying vitality of local privilege. For most Europeans these frontiers were as significant as their international counterparts. The two types were difficult to distinguish on seventeenth-century maps. This mental world was not to change appreciably until the impetus that the French Revolution gave to nationalism altered the European political consciousness.

NOTES

1 I would like to thank David Aldridge, Matthew Anderson, Peter Barber, Sarah Bendall, Hugh Dunthorne, P.D.A. Harvey, Josef Konvitz, Stewart Oakley, Johannes Pallière and John Stoye for their comments on earlier drafts. I am grateful to the British Academy and to Durham University for their support of my research and to Janet Forster for her assistance.
2 I would like to thank John Stoye for lending me a copy of his then unpublished paper.
3 These maps are displayed at the Musée de la Compagnie des Indes at Port-Louis in Brittany.
4 I owe this point to Professor Norman Housley.

REFERENCES

Abou-El-Haj, R.A. (1969) 'The Formal Closure of the Ottoman Frontier in Europe: 1699–1703', *Journal of the American Oriental Society*, 89.
Ade Ajayi, A.F. and Crowder, M. (eds) (1985) *Historical Atlas of Africa*, Cambridge.
Allies, P. (1980) *L'Invention du territoire*, Grenoble.
Almagia, R. (1929) *Monumenta Italiae Cartographica*.
Andrews, J. (1985) *Plantation Acres*, Belfast.
Archives du Ministère des Affaires Etrangères, Correspondance Politique, Sardaigne 231.
Archives du Ministère des Affaires Etrangères, Mémoires et Documents, France et Divers Etats, vol. 36.

Atkin, M. (1980) *Russia and Iran, 1780–1828*, Minneapolis.

Atlas Geographus (1740), London.

Atlas of Pennsylvania, The (1989), Philadelphia.

Baillou, J. (ed.) (1984) *Histoire de l'Administration Française. Les Affaires Etrangères et le Corps Diplomatique Français*, Paris.

Barber, P. (1990) 'Necessary and Ornamental: Map Use in England under the Later Stuarts, 1660–1714', *Eighteenth-Century Life*, 14.

—— (1991) 'Henry VIII and Mapmaking', in D. Starkey (ed.), *Henry VIII. A European Court in England*, London.

Barber, P. and Black, J.M., 'Maps and the Complexities of Eighteenth-Century Europe's Territorial Divisions: Holstein in 1762', unpublished paper.

Bartlett, R. and Mackay, A. (1990) *Medieval Frontier Societies*, Oxford.

Bassett, D.K. (1971) *British Trade and Policy in Indonesia and Malaysia in the Late Eighteenth Century*, Hull.

Bayly, C.A. (1989) *Imperial Meridian: The British Empire and the World 1780–1830*, London.

Bédarida, H. (1928) *Parme et la France de 1748 à 1789*, Paris.

Bedford, Duke of (1749), Secretary of State for the Southern Department, to Benjamin Keene, envoy in Spain, 24 April, 5 June, Public Record Office, State Papers 94/135.

Bemis, S.F. (1960) *Pinckney's Treaty: America's Advantage from Europe's Distress*, rev. edn, New Haven.

Bendall, A.S. (1992) *Maps, Land and Society. A History, with a Carto-Bibliography, of Cambridgeshire Estate Maps, 1600–1836*, Cambridge.

Black, J.D. (ed.) (1970–5) *The Blathwayt Atlas*, 2 vols, Providence.

Black, J.M. (1985) *British Foreign Policy in the Age of Walpole*, Edinburgh.

—— (1986) 'Fresh Light on Ministerial Patronage of Eighteenth-Century Pamphlets', *Publishing History*, 19.

—— (ed.) (1987) *The Origins of War in Early Modern Europe*, Edinburgh.

—— (1988) 'French Foreign Policy in the Age of Fleury Reassessed', *English Historical Review*, 103.

—— (1990a) 'Anglo-French Relations in the Mid-Eighteenth Century 1740–1756', *Francia*, 17.

—— (1990b) *The Rise of the European Powers 1679–1793*, London.

—— (1990c) *Eighteenth Century Europe 1700–1789*, London.

—— (1992) 'Maps and Chaps: The Historical Atlas. A Perspective from 1992', *Storia della Storiografia*, 21.

—— (1994) *British Foreign Policy in an Age of Revolutions 1783–1793*, Cambridge.

Blanning, T.C.W. (1986) *The Origins of the French Revolutionary Wars*, Harlow.

Bonenfant, P. (1953) 'A propos des limites médievales', in *Hommage à Lucien Febvre: Eventail de l'histoire vivant*, 2 vols, vol. I, Paris.

Bonnac (1755) to Rouillé, 21 February, Archives du Ministère des Affaires Etrangères, Correspondance Politique, Hollande 488.

Bonney, R. (1971) *Kedah 1771–1821. The Search for Security and Independence*, Oxford.

Bourguet, M.N. (1988) *Déchiffrer la France. La statistique départementale à l'époque napoléonienne*, Paris.

Brown, L. (1941) *Jean Dominique Cassini and His World Map of 1696*, Ann Arbor.

Bruchet, M. (1896) *Notice sur l'ancien cadastre de Savoie*, Annecy.

Buisseret, D. (1982) 'Cartography and Power in the Seventeenth Century', *Proceedings of the Western Society for French Historical Studies*, 10.

Buisseret, D. (1985) 'The Cartographic Definition of France's Eastern Frontier in the Early Seventeenth Century', *Imago Mundi*, 36.

—— (1992) (ed.) *Monarchs, Ministers and Maps. The Emergence of Cartography as a Tool of Government in Early Modern Europe*, Chicago.

Bussy (1755) to Rouillé, 29 July, Archives du Ministère des Affaires Etrangères, Correspondance Politique, Brunswick-Hanovre 52.

Cambridge History of Iran (1991) VII, Cambridge.

Cashin, E.J. (1992) *Lachlan McGillivray, Indian Trader. The Shaping of the Southern Colonial Frontier*, Athens, Georgia.

Chambers, D. and Pullan, B. (eds) (1992) *Venice. A Documentary History 1450–1630*, Oxford.

Chauvelin (1728), Morville's successor, to Richelieu, 12 February, Bibliothèque Victor Cousin, Fonds de Richelieu 32.

Chavigny (1727), Groffey's successor, to Richelieu, envoy at Vienna, 7 January, Bibliothèque Victor Cousin, Fonds de Richelieu 32.

Christian IV, Duke of Zweibrücken, to his Parisian envoy, Wernike, Munich, Bayerisches Hauptstaatsarchiv, Abteilung II, Gesandtschaften Paris 216.

Clayton, T.R. (1981) 'The Duke of Newcastle, the Earl of Halifax, and the American Origins of the Seven Years War', *Historical Journal*, 24.

Cockran, D.H. (1967) *The Creek Frontier, 1540–1783*, Norman.

Cole Harris, R. (ed.) (1987) *Historical Atlas of Canada. I: From the Beginning to 1800*, Toronto.

Conference of 25 December 1734, Public Record Office, State Papers 84/337.

Cook, W.L. (1973) *Flood Tide of Empire: Spain and the Pacific Northwest, 1543–1819*, New Haven.

Crane, V.W. (1929) *The Southern Frontier, 1670–1732*, Ann Arbor.

Cumming, W.P. (1962) *The Southeast in Early Maps*, Chapel Hill.

Daily Universal Register (1786), 11 August.

d'Albissin, N.G. (1966) 'Propos sur la frontière', *Revue Historique de Droit Français et Etranger*, 47.

—— (1979) *Genèse de la frontière franco-belge. Les Variations des limites septentrionales de la France de 1659 à 1789*, Paris.

Darby, H.C. and Fullard, H. (eds) (1970) *The New Cambridge Modern History Atlas*, Cambridge.

de Dainville, F. (1970) 'Cartes et contestations au XVe siècle', *Imago Mundi*, 24.

de Lapradelle, P. (1928) *La Frontière: Etude de droit international*, Paris.

de Silly, Marquis (1727) to Richelieu, envoy at Vienna, 14 March, Bibliothèque Victor Cousin, Fonds de Richelieu, 33.

DeVorsey Jr, L. (1966) *The Indian Boundary in the Southern Colonies, 1763–1775*, Chapel Hill.

Dickson, P.G.M. (1991) 'Joseph II's Hungarian Land Survey', *English Historical Review*, 106.

Downes, R.C. (1970) *Evolution of Ohio County Boundaries*, Columbus, Ohio.

Ekberg, C.J. (1979) *The Failure of Louis XIV's Dutch War*, Chapel Hill.

Finch, Edward (1741) to Earl of Harrington, 8 December, Public Record Office, State Papers 91/29.

Fisher, A.W. (1970) *The Russian Annexation of the Crimea 1772–1783*, Cambridge.

Forbes, E.G. (1974) *The Birth of Scientific Navigation: The Solving in the Eighteenth Century of the Problem of Finding Longitude at Sea*, Greenwich.

Freeman-Grenville, G.S.P. (1991) *The New Atlas of African History*, New York.

Gallois, L. (1909) 'L'Académie des sciences et les origines de la carte de Cassini', *Annales de Géographie*, 18.

Gammon, S.R. (1973) *Statesman and Schemer. William, First Lord Paget Tudor Minister*, Newton Abbot.

Gentleman's Magazine (1791) 61.

George III (1792) to Lord Grenville, Foreign Secretary, 26 September, 37–9.

Grantham, Lord (1775), British envoy in Madrid, to Horace St Paul, Secretary of Embassy at Paris, 8 May, 20 November, British Library, Add. 24177.

Greengrass, M. (ed.) (1991) *Conquest and Coalescence. The Shaping of the State in Early Modern Europe*, London.

Grendi, E. (1986) 'La pratica dei confini: Mioglia contro Sassello, 1715–1745', *Quaderni Storici*, 21.

Grenville, Lord (1792), Foreign Secretary, to George III, 25 September, British Library, Add. 58857.

Griffin, T. and McCaskill, M. (eds) (1986) *Atlas of South Australia*, Adelaide.

Guyot, E. (1955) *Histoire de la determination des longitudes*, La Chaux-de-Fonds.

Hall, C.D. (1992) *British Strategy in the Napoleonic War 1803–15*, Manchester.

Hansen, R. 'Die Nordgrenze Deutschlands im Lauf der Geschichte', *Grenzfriedenshefte*, 1/90.

Harley, J.B. (1987) 'Maps, Knowledge and Power', in D. Cosgrove and S.J. Daniels (eds), *The Iconography of Landscape*, Cambridge.

—— (1988) 'Silences and Secrecy: The Hidden Agenda of Cartography in Early Modern Europe', *Imago Mundi*, 40.

—— (1992) 'The Society and the Surveys of English Counties, 1759–1809', in D.G.C. Allan and J.L. Abbott (eds), *The Virtuoso Tribe of Arts and Sciences. Studies in the Eighteenth-Century Work and Membership of the London Society of Arts*, Athens, Georgia.

Harley, J.B. and Woodward, D. (eds) (1987) *History of Cartography*, vol. I, Chicago.

Harsin, P. (1927) *Les Relations extérieures de la Principauté de Liège*.

Haslam, G. (1991) 'Patronising the Plotters: The Advent of Systematic Estate Mapping', in K. Barker and R.J.P. Kain (eds), *Maps and History in South-West England*, Exeter.

Helmfrid, S. (1990) 'Five Centuries of Sweden on Maps', in U. Sporrong and H.F. Wennstrom (eds), *Maps and Mapping*, Stockholm.

Hetherington, P. (1991) 'Anglo-Scottish Borders', *Boundary Bulletin*, 2.

Hodson, Y. (1987) 'The Military Influence on the Official Mapping of Britain in the Eighteenth Century', *IMCOS Journal*, 27.

Holdernesse, Earl of (1749), envoy at The Hague, to Duke of Newcastle, Secretary of State for the Northern Department, 2 December, Public Record

Office, State Papers 84/449.
Howse, D. (1980) *Greenwich Time and the Discovery of the Longitude*, New York.
—— (ed.) (1990) *Background to Discovery. Pacific Exploration from Dampier to Cook*, Berkeley.
Inalcik, H. (1954) 'Ottoman Methods of Conquest', *Studia Islamica*, 2.
Instructions (1716) to Marquis d'Entremont, 1 May, in A. Manno, E. Vayra and E. Ferrero (eds) (1886–91), *Relazioni diplomatiche della monarchia di Savoia dalla prima alla seconda restaurazione 1559–1814: Francia, Periodo III, 1713–1719*, II, 148.
Jenkinson, Charles (1785), Anglo-Spanish treaty of 8/18 July 1670, *A Collection of all the Treaties . . . between Great Britain and Other Powers, from . . . 1648, to . . . 1783*, 3 vols, London.
Jewsbury, G. (1976) *The Russian Annexation of Bessarabia: 1774–1828*, Boulder.
Kaunitz (1789), Austrian Chancellor, to Noailles, French envoy in Vienna, and Llano, Spanish envoy, 29 September, 18 December, Archives du Ministère des Affaires Etrangères, Correspondance Politique, Autriche 357.
Keay, J. (1991) *The Honourable Company. A History of the English East India Company*, London.
Keene, Benjamin (1749), envoy in Spain, to Duke of Bedford, Secretary of State for the Southern Department, 21 May, 29 June, Public Record Office, State Papers 94/135.
Keith, Robert Murray (1791) to Lord Grenville, Foreign Secretary, 5 August, Public Record Office, Foreign Office 7/27.
Keith, Robert (1752) to Duke of Newcastle, 23 August, Public Record Office, State Papers 80/190.
King, P. (1987) *Charlemagne. Translated Sources*, Kendal.
Kirby, D. (1990) *Northern Europe in the Early Modern Period. The Baltic World 1492–1772*, Harlow.
Klang, D.M. (1977) *Tax Reform in Eighteenth Century Lombardy*, Columbia.
Konvitz, J.W. (1987) *Cartography in France, 1660–1848. Science, Engineering and Statecraft*, Chicago.
—— (1990) 'The Nation-State, Paris and Cartography in Eighteenth- and Nineteenth-Century France', *Journal of Historical Geography*, 16.
Lang, D.M. (1957) *The Last Years of the Georgian Monarchy 1658–1832*, New York.
Lanning, J.T. (1936) *The Diplomatic History of Georgia: A Study of the Epoch of Jenkins' Ear*, Chapel Hill.
Literary Magazine (1756) 15 October.
Livet, G. (ed.) (1966) *Recueil des instructions données aux ambassadeurs et ministres de France depuis les traités de Westphalie jusqu'à la Révolution Française. L'Electorat de Trèves*, Paris cxix–cxx, cxxxiii–cxl.
Louis XV (1745) to Vaulgrenant, 21 October, Paris, Bibliothèque Victor Cousin, Fonds de Richelieu 40, f. 77.
Ludwig, E. (1726), Duke of Württemberg, to -, 23 May, Paris, Bibliothèque Victor Cousin, Fonds de Richelieu 31.
Ludwig, T. (1898) *Die deutschen Reichsstände in Elsass und der Ausbruch der Revolutionskriege*, Strasburg.

Magocsi, P.R. (1985) *Ukraine: A Historical Atlas,* Toronto.
Malmesbury, Third Earl of (ed.) (1844) *Diaries and Correspondence of James Harris, First Earl of Malmesbury,* 4 vols, vol. II, London.
Manno, A., Vayra, E. and Ferrero, E. (eds) (1886–91) *Relazioni diplomatiche della monarchia di Savoia dalla prima alla seconda restaurazione 1559–1814: Francia, Periodo III, 1713–1719,* 3 vols, Turin.
Mason, A.S. (1990) *Essex on the Map. The Eighteenth-Century Land Surveyors of Essex,* Chelmsford.
Mirepoix (1755), French envoy in London, to Rouillé, French foreign minister, 16 January, 8 March, Paris, Archives du Ministère des Affaires Etrangères, Correspondance Politique 438.
Montaigu (1743), French envoy in Venice, to Amelot, French foreign minister, 10 August, Bibliothèque Nationale, Nouvelles Acquisitions Françaises 14904.
Montmorin (1790) to Noailles, 2 April, Paris, Archives du Ministère des Affaires Etrangères, Correspondance Politique, Autriche 359.
Moore, J.N. (1988) 'Scottish Cartography in the Later Stuart Era, 1660–1714', *Scottish Tradition,* 14.
Morville (1725), French foreign minister, to Groffey, envoy at the Imperial Diet, 18 October, Paris, Bibliothèque Nationale, Manuscrits Français 10681.
Murphy, O.T. (1982) *Charles Gravier. Comte de Vergennes,* Albany.
New Cambridge Modern History XIV. Atlas (1970), Cambridge.
Newcastle, Duke of (1749), Secretary of State for the Northern Department, to Earl of Holdernesse, envoy at The Hague, 22 December, Public Record Office, State Papers 84/449.
Noailles (1789), French envoy in Vienna, to Montmorin, French foreign envoy, 30 May, Paris, Archives du Ministère des Affaires Etrangères, Correspondance Politique, Autriche 358.
—— (1790) to Montmorin, French foreign envoy, 7 April, Archives du Ministère des Affaires Etrangères, Correspondance Politique, Autriche 359.
Noailles and Llano (1789), Spanish envoy, to Kaunitz, Austrian Chancellor, 23 September, Archives du Ministère des Affaires Etrangères, Correspondance Politique, Autriche 357.
Noel, J.F. (1946) 'Les problèmes des frontières entre la France et l'Empire dans la seconde moitié du XVIIIᵉ siècle', *Revue Historique,* 235.
Nordman, D. and Revel, J. (1989) 'La formation de l'espace français', in J. Revel (ed.), *Histoire de la France,* vol. I: *L'Espace français,* Paris.
O'Donaghue (1977) *William Roy (1726–1790): Pioneer of the Ordnance Survey,* London.
Oliver, R. (1991) 'The Ordnance Survey in South-West England', in K. Barker and R.J.P. Kain (eds), *Maps and History in South-West England,* Exeter.
Olson, R.W. (1975) *The Siege of Mosul,* Bloomington.
Ozouf-Marignier, M.V. (1984) 'Territoire géométrique et centralité urbaine: Le découpage de la France en départements, 1789–1790', *Annales de la Recherche Urbaine,* 22.
—— (1986) 'De l'universalisme constituant aux intérêts locaux: Le débat sur la formation des départements en France (1789–1790)', *Annales,* 41.
—— (1989) *La Formation des départements; La Représentation du territoire française à la fin du XVIIIᵉ siècle,* Paris.

Pallière, J. (1979) 'Un grand méconnu du XVIIIᵉ siècle: Pierre Bourcet (1700–1780)', *Revue Historique des Armées.*

—— (1984) *Histoire de l'Administration Française. Les Affaires Etrangères et le Corps Diplomatique Français,* 2 vols, vol. I, Paris.

—— (1985a) 'Le maître savoyard de la cartographie, Antoine Durieu (1703–1777)', *Actes du 109ᵉ Congrès National des Sociétés Savantes, Dijon, 2–6 avril 1984. Section de géographie,* Paris.

—— (1985b) 'Les cartes de 1760–1764 et la frontière Franco-Sarde', *Actes du 110ᵉ Congrès National des Sociétés Savantes, Montpellier 1985. Section de géographie,* Paris.

—— (1986) 'Le traité du 24 mars 1760 et les nouvelles frontières de la Savoie', Société Savoienne d'Histoire et d'Archéologie, *Frontières de Savoie, L'Histoire en Savoie.*

Pease, T.C. (ed.) (1936) *Anglo-French Boundary Disputes in the West, 1749–1763,* Springfield, Illinois.

Pelletiev, M. (1984) 'La Martinique et La Guadeloupe au lendemain du traité de Paris, l'œuvre des ingénieurs géographes', *Chronique d'Histoire Maritime,* 9.

Penfold, P.A. (ed.) (1974) *Maps and Plans in the Public Record Office, II: America and the West Indies,* London.

Perrot, J.C. (1977) *L'Age d'or de la statistique régionale française (an IV–1804),* Paris.

Perrot, J.C. and Woolf, S. (1984) *State and Statistics in France 1789–1815.*

Poyntz (1746) to Edward Weston, 18 October, Farmington Connecticut, Lewis Walpole Library, Weston papers, vol. 18.

Praslin (1763), French foreign minister, to Châtelet, French envoy in Vienna, 16 July, Archives du Ministère des Affaires Etrangères, Correspondance Politique, Autriche 295.

Puysieulx (1748) to Duke of Richelieu, French commander at Genoa, 22 July, Paris, Archives Nationales, KK 1372.

—— (1750) to Valory, 29 August, Archives du Ministère des Affaires Etrangères, Correspondance Politique, Brunswick-Hanovre 50.

Quazza, G. (1957) *Le riforme in Piemonte nella prima metà del '700,* Modena.

Ranum, O. (1991) 'Louis XV and the Price of Pacific Inclination', *International History Review,* 13.

Rashed, Z.E. (1951) *The Peace of Paris 1763,* Liverpool.

Ravenhill, W. (1991) 'The South West in the Eighteenth-Century Re-mapping of England', in K. Barker and R.J.P. Kain (eds), *Maps and History in South-West England,* Exeter.

Rebenac (1688), French envoy in Madrid, to Louis XIV, 9 September, Archives du Ministère des Affaires Etrangères, Correspondance Politique, Espagne 75.

Reese, A. (1988) *Europäische Hegemonie und France d'outre-mer. Koloniale Fragen in der französischen Aussenpolitik 1700–1763,* Stuttgart.

Reitan, E.A. (1985) 'Expanding Horizons: Maps in the *Gentleman's Magazine,* 1731–1754', *Imago Mundi,* 37.

Richard, J. (1948) 'Enclaves royales et limites des Provinces', *Annales de Bourgogne,* 20.

Ritcheson, C.R. (1969) *Aftermath of Revolution: British Policy towards the United States 1782–1795,* Dallas.

Robinson, Thomas (1733), envoy in Vienna, to Weston, 27 May, Public Record Office, State Papers 80/96.
Roider, K.A. (1982) *Austria's Eastern Question 1700–1790*, Princeton.
Sack, R.D. (1986) *Human Territoriality: Its Theory and History*, Cambridge.
Sackville, (1758) Lord George to Holdernesse, 1 October, British Library, Eg. 334, 75.
Sahlen (1721) to Sir Luke Schaub, George I's envoy in Paris, 6 October, New York, Public Library, Hardwicke papers vol. 56.
Sahlins, P. (1989) *Boundaries: The Making of France and Spain in the Pyrenees*, Berkeley.
—— (1990) 'Natural Frontiers Revisited: France's Boundaries since the Seventeenth Century', *American Historical Review*, 95.
Sandwich (1748) to Duke of Newcastle, 28 April, Public Record Office, State Papers 84/433.
Savelle, M. (1940) *The Diplomatic History of the Canadian Boundary, 1749–1763*, New Haven.
Seymour, W.A. (1980) *A History of the Ordnance Survey*, Folkestone.
Shelby, L.R. (1967) *John Rogers. Tudor Military Engineer*, Oxford.
Shirley, R. (1984) *The Mapping of the World, 1472–1700*, Watchung, New Jersey.
Solon, P. (1982) 'Frontiers and Boundaries: French Cartography and the Limitation of Bourbon Ambition in Seventeenth-Century France', *Proceedings of the Western Society for French Historical Studies*, 10.
Somme, A. (ed.) (1968) *A Geography of Norden, Denmark, Finland, Iceland, Norway, Sweden*.
Sosin, J.M. (1967) *The Revolutionary Frontier, 1763–1783*, New York.
Stoye, J. (1993) 'The Treaty of Carlowitz and the New Frontier', in *Marsigli's Europe, 1680–1730. The Life and Times of Luigi Ferdinando Marsigli, Soldier and Virtuoso*, New Haven.
Stuart, R.C. (1988) *United States Expansionism and British North America, 1775–1871*, Chapel Hill.
Subtelny, O. (1986) *Domination of Eastern Europe. Native Nobilities and Foreign Absolutism, 1500–1715*, Gloucester.
Szarka, A.S. (1975) 'Portugal, France, and the Coming of the War of the Spanish Succession', Ph.D. thesis, Ohio State University.
Thomas, H.M. (1992) *A Catalogue of Glamorgan Estate Maps*, Cardiff.
Topolski, J. (1969) 'The Polish Prussian Frontier during the Period of the First Partition (1772–1777)', *Polish Western Affairs*, 10.
Torcy (1712) to St John, 28 July, Public Record Office, State Papers 78/154.
Tyacke, S. (ed.) (1983) *English Map-Making 1500–1650*, London.
Vienna, Haus- Hof-, und Staatsarchiv, Nachlass Fonseca: includes a volume on the Liège frontier 1718–30 and another on Saint-Hubert 1716–25.
Waeber, P. (1974) *La Formation du Canton de Genève*, Geneva.
Watelet, M. (1990/4) 'La cartographie topographique militaire des Alliés en France et en Belgique (1815–1818)', *Bulletin Trimestriel du Crédit Communal de Belgique*, 174.
Webb, P. (1975) 'The Naval Aspects of the Nootka Sound Crisis', *Mariner's Mirror*.
Westminster Journal (1749) 25 March.

Whitaker, A.P. (1927) *The Spanish–American Frontier, 1783–1795: The Westward Movement and the Spanish Retreat in the Mississippi Valley*, Boston.

Whitworth, Francis and Polwarth, Lord (1723), envoys at the Congress, to Lord Carteret, Secretary of State, 4 November, British Library, Add. 32792.

Whitworth (1715) to Sir Luke Schaub, George I's envoy in Paris, 9 July, New York, Public Library, Hardwicke papers vol. 42.

Woolf, S. (1991) *Napoleon's Integration of Europe*, London.

Wright, J.L. (1975) *Britain and the American Frontier, 1783–1815*, Athens, Georgia.

Yorke, Joseph (1749a) to his father, Lord Chancellor Hardwicke, 5 April, British Library, Add. 35355.

—— (1749b) to Hardwicke, 27 August, British Library, Add. 35355.

2

THE INNER-GERMAN BORDER

Consequences of its establishment and abolition

Hanns Buchholz

The consequences of state borders have to be considered on two levels. First, borders show the area of the state's direct political power. They secure and they limit, for example, the legal system, the economic system (tax, currency and so on) and the sphere of a specific way of life. In this way the state borders divide a region into areas of differing political, administrative, economic and social forms of organization. In the case of the two German states this meant, for instance, on the eastern side a highly centralized government system, hierarchically organized with a strong commanding character, and on the western side a federal system with self-governing entities at different administrative levels. Or, with respect to agriculture: huge co-operatives in Eastern Germany, individual private farms in Western Germany. Or with respect to property ownership: state-owned in the East, mainly private in the West.

Second, state borders create special border areas with specific characteristics caused by the immediate effects of the boundary. The restricted possibilities of crossing the boundary are most significant and this means the reduced accessibility of locations on the other side of the border. Additionally, the border area is sometimes covered by special laws for security reasons. Of course, these 'immediate effects' are superimposed by the respective political system.

This chapter will highlight only the second consequences of the inner-German border, with the proviso that the time since the opening of the border has been too short to present the final results of this border-zone research. Geographical research on the inner-German border zone is becoming most attractive due to the fact that this state border was established arbitrarily in the centre of a more or less equally structured region, that this border existed for more than forty years, and that it then suddenly disappeared without any military or similar actions which would have affected the border area in particular. This situation provides the opportunity for very deep insights into the spatial (and other) effects of a border.

THE ESTABLISHMENT OF THE INNER-GERMAN BORDER

At the Conference of Teheran (1943) the three Allies – the United Kingdom, the United States of America and the Soviet Union – established the so-called 'European Advisory Commission' (EAC) in order to outline proposals for the partition of Germany into three occupation zones. The EAC concluded its consultations with a 'Protocol regarding the occupation zones in Germany and the Administration of Greater Berlin'. In this 'protocol' the boundary between the Soviet Occupational Zone on the one side and the occupation zones of the Western allies was defined as the line of the western boundaries of the German states (Länder) of Mecklenburg in the north, the Prussian province of Saxony (not to be confused with the state of Saxony) in the centre, and of Thuringia in the south.

This general delimitation by using historical boundaries was straightened in many places for practical reasons soon after the war, but this is not a subject of this chapter. It should be considered that the old German borders of states and provinces had been open for a very long time, at least from 1834 onward, because of the foundation of the German Customs Union (Zollverein), or from 1871 onward with the formation of the second German Empire. The boundaries had been reduced to merely an administrative status.

With the birth of the Federal Republic of Germany on 23 May 1949, and the formation of the German Democratic Republic (GDR) on 7 October of the same year, the 'demarcation line' between the occupation zones was given a new quality as a real state border, although it was not formally accepted by the Federal Republic of Germany.

At the beginning the border was kept open for a variety of cross-border connections, including a large amount of illegal border traffic. Before May 1952 there were thousands of legal commuters between the two German states; only then was this type of crossing abolished. In 1961, parallel to the construction of the Berlin Wall, the inner-German border was sealed almost hermetically by the GDR government. Step by step, the eastern side of the boundary assumed the character of a fortification that was 1,393 kilometres long. In 1951 a fence of barbed wire, 1.20 metres high, was built, and a series of zones was developed adjacent to the boundary: first, a 10-metre Controlled Zone where the use of firearms was permitted for GDR border troops; second, a 500-metre Security Zone, which could be entered solely by GDR subjects with a special permit, and only during daylight (restaurants, cinemas, pensions, health resorts and so on were closed down); and, third, further inland there followed a 5-kilometre Restricted Zone, where the residents were given a special stamp on their identity card for unlimited permission to enter and to leave the zone, but only for the 'inland' part of the GDR (that is, this did not permit visits to neighbouring villages in the Restricted Zone). All these installations and arrangements were not able to limit the growing number of refugees from the GDR to the Federal Republic of Germany: between 1949 and 1961 about 200,000–300,000 residents each year left the GDR illegally.

In 1961 the barbed-wire fence was raised to 1.50 metres, and the deployment of mines in wooden boxes was begun. In 1966 the barbed-wire fence was substituted by metal plates 2.40 metres high, accompanied by the deployment of plastic mines. From 1970 onward the metal-plate fence was raised to 3–4 metres with spring-guns attached, everything inland-oriented contrary to any usual state border.

In 1980 there were 1,277 kilometres of metal fence, 232 kilometres of mines, 411 kilometres of spring-guns and 832 kilometres of blocking trench. On the GDR side about 48,000 border troops guarded the boundary; on the West German side about 22,300 border police and customs officers were on duty. There was interruption to 10 main railway lines, 24 secondary railway lines, 23 motorways (Autobahn) or national roads, 140 regional roads and thousands of smaller roads and paths. By 1966 only six railway lines had been left open, three motorways, one regional road and two waterways. During the following years the GDR government agreed to open up some more border crossing-points in return for economic advantages offered by West Germany.

Despite these fortifications, in general the inner-German border has never been as tight as, for example, the border between South and North Korea. During all the years of its existence, legal opportunities to cross the border were always provided for, even if strongly restricted by bureaucratic complications. Most of the visitors crossing the border travelled from West to East. In 1980 about 3 million West Germans visited the GDR, and about 1.4 million pensioners from the GDR plus about 40,000–50,000 non-pensioners visited West Germany. At almost no time did the GDR authorities give complete families permission to visit the West because they suspected they would never come back. Parallel to this, mail and telephone connections were provided, even if all letters and parcels were checked and telephone calls monitored.

In addition it should be mentioned that special arrangements for the four Allies permitted traffic connections between the Federal Republic of Germany and West Berlin along reserved corridors by car, by rail and by air. Only land communication was interrupted by the so-called 'Berlin blockade', 24/30 June 1948–12 May 1949.

CONSEQUENCES OF THE BOUNDARY

Whereas the border was open for officially sanctioned inter-state communications, it was hermetically sealed for virtually all 'low-level' activities. The border areas on both sides became very distant locations. They developed into extremely peripheral regions with all the typical structural changes: absence of investment, decrease of employment, out-migration of the younger and economically active people in particular, lack of educational and vocational training opportunities – especially regarding their variety, over-ageing of the remaining population, down-grading of the infrastructure and the extensification of agriculture.

The two German states reacted to this development in different ways: in the Federal Republic the process of degrading was limited by public support instruments, and from 1971 onward by the so-called 'Zonenrandförderung', a special programme supporting the belt along the border. Aid was given to all counties and cities if more than 50 per cent of their area, or if more than 50 per cent of the population, was located less than 40 kilometres from the inner-German border (including the respective belt along the Czechoslovakian boundary and the Baltic Sea coast). Large amounts of money were spent on reconstructing and modernizing villages and towns, on improving the technical infrastructure (roads, fresh water and sewerage systems and so on), on building new schools, sports grounds, swimming pools, city halls, village community houses and on supporting the establishment of factories and new enterprises. On the whole these measures have prevented the West German border area from total decline, despite the fact that the creation of new employment opportunities has not been very successful: very often the investors used the publicly promoted cheap credits and tax reductions only for the period of time that was needed to avoid repayment, and then closed down their undertakings.

In many other cases branches of more centrally located companies were opened up in the West German border area in periods of favourable business conditions and ceased operation in times of declining trends. However, the renewed or newly constructed infrastructure still exists, so that at least the general framework for the living conditions in this area has developed positively.

Some parts of the West German border area profited economically from the establishment of the inner-German border. For example, the 'Coburg Land' south of the Thuringian Forest has benefited from the movement of entrepreneurs and skilled workers from the Soviet Occupational Zone, or later the GDR, in order to settle on the western side of the border and to continue with their former business. In this way the Coburg Land experienced a favourable development, additionally assisted by the fact that the increasingly strict border controls reduced competition from east of the border.

The border zone on the eastern side in the GDR generally declined economically. The main difference between this zone and the corresponding zone in the West derives from the fortification measures in the East. The establishment of the Security Zone and the Restricted Zone by the GDR government created a buffer zone in the borderland. In official statements this buffer zone was directed against assumed aggression from the West, but in reality it was established to prevent refugees from the East from leaving their country. Therefore many people who were politically 'unstable' – 8,422 persons in 1952 and 3,640 persons in 1961 (information from the Federal Ministry of Justice, 30 March 1992) – were removed and resettled elsewhere. Many younger people left the border area in order to find better employment and generally more attractive living conditions elsewhere in the GDR. On the one hand, the GDR authorities did nothing to improve the economic base or to modernize the infrastructure in the border area. On the other hand, they feared the depletion of population living in the area. Therefore all the residents of this zone received a 15 per cent supplement to their income, and credits for private single-family houses were granted much more liberally than in the GDR proper (Decree of the Ministry of State Security, 1 June 1952). Finally, these extra costs together with many difficulties resulting from the complicated accessibility of this area led to an eventual reduction in the area of the Restricted Zone in 1972.

HANNS BUCHHOLZ

CONSEQUENCES OF THE ABOLITION OF THE BORDER

On 9 November 1989 the state border was suddenly opened. Very soon the border installations were demolished and removed. People from both sides could cross the border line without any official restrictions.

However, after the abolition of the dividing barrier there were, and still are, decisive differences and difficulties so far as a positive and common development is concerned. It is not only a question of repairing unmaintained infrastructure, for example destroyed bridges crossing the border creeks, or roads intersected by the former border. The main problem derives from the fact that none of this infrastructure was modernized or enlarged after the 1940s. The roads are too small for the demands of modern traffic, and the interrupted railway lines were demolished on the GDR side, so that a completely new and comprehensive system has to be re-established. Similar difficulties exist for water-pipes, sewerage systems and the electricity supply.

Another problem results from the differing tenureship systems between West and East. Whereas proprietary rights of GDR residents in Western Germany remained untouched (West German leaseholders had to pay a rent), the land of West German farmers on the eastern side of the boundary was expropriated and given to one of the large co-operatives. It is a very complicated process to get the land back, despite the fact that the Federal government has decided to follow the principle of 'retrocession before compensation'. Very often the plots are fully integrated in a very different farming system and sometimes it is even difficult to identify the old pieces of land because the GDR authorities did not keep detailed land registers. Within the 10-metre Controlled Zone and in some other locations of strategic importance the farmland was transferred to the GDR military forces or to other state authorities, often with no written contract. The state and the border troops (belonging to the GDR's Ministry of State Security) held more-or-less absolute power, and the co-operatives had no objections because every ceded hectare of arable land diminished the harvest quota demanded by the state planning commission.

Further inland there is often the chance of some arrangement, because in many cases the West German landlords are not so interested in becoming farmers in the 'new states'. Close to the former boundary the situation is different for the farmer from the West who is making the claim often lives just on the other side of the boundary line and is therefore eager to add his eastern plots direct to his West German property.

Another interesting and problematic phenomenon appears at the former border: in the former GDR the number of real farmers has diminished remarkably because the huge co-operatives needed lots of specialists (chemists, veterinarians, tractor drivers, accountants and so on) but few farmers. Therefore the agrarian structure in the new states will never return to the traditional system of smallholders. Company farms or new types of private co-operatives will become symptomatic of the agriculture in the former GDR – and at the former inner-German border these two very different types of farming are located side by side.

A decisive change has occurred to commuting systems because of the fact that in the east more than a million employees became jobless after the opening of the border and after the change from a socialist system to a market-orientated system. There has been a sharp rise in the number of commuters from the eastern border area to jobs in the West, as well as a long-term movement of Germans from East to West. Another change has taken place regarding the central places and their functional areas. No longer does any border limit the mobility of the people, so that the people may decide themselves where to buy goods or where to use services. Because of the higher standard of services, goods and shops on the western side the majority of residents of the eastern border area regularly travel to central towns in the West. Consequently the retail and service centres in the West have increased enormously. Buying power is moving from East to West so that the recovery of the eastern border area is additionally handicapped.

The overall consequence of the abolition of the inner-German border is the transformation of the border zone from an extreme periphery to the very centre of Germany and Europe. This does not mean that the whole 1,393-kilometre-long border belt will become a boom area. Some parts will remain economically poor and removed from modern developments. The border zone also offers opportunities to develop the natural environment, since the border area was almost untouched for several decades and therefore offers a rich variety of plants and animals. However, in general the accessibility of this border zone has increased tremendously, so that several sections will change their character in the near future.

This spatial re-evaluation does not apply only to the border zone. Large parts of Germany – if not of Europe – are changing with regard to their locational value because of the abolition of the inner-German border; and this will influence the border area itself. Perhaps the most obvious change will be in the main communication network. Up to November 1989 the main traffic directions were from north to south in

East Germany as well as in West Germany; the east–west directions were considerably disturbed. Now we can observe a change in these relations, and this will provide great opportunities for the locations where one of the new axes crosses the former inner-German border. In this way the Central European Axis between London/Brussels/ Rotterdam–Hanover–Berlin–Moscow/Kaliningrad/the Baltic states will become one of the most significant European development axes, and the former borderland is located right in the centre of this axis.

Consequently, this formerly quiet, unchanging, depopulated, over-aged and economically backward area will become the target of dynamic developments. The spatial concepts of both sides have to be re-written, but the 'mental map' of planners and politicians from both sides has not always adapted to the new situation. Of course, all areas of European states undergo continuous change; but the area close to the former state border between the GDR and the old Federal Republic of Germany will experience an even greater and deeper change.

REFERENCES

Bode, V. (in print) *Die Raumbedeutsamkeit einer Staatsgrenze. Die Auswirkungen der ehemaligen innerdeutschen Grenze auf den grenznahen Raum Sachsen-Anhalts.*

Boesler, K.A. (1985) 'Das Zonenrandgebiet. Eine Einführung in die aktuellen Probleme seiner Struktur und Entwicklung', *Geographische Rundschau*, 37: 380–4.

Frische, W. (1979) 'Duderstadt. Probleme einer Kleinstadt an der inner-deutschen Grenze', *Geographische Rundschau*, 31: 187–94.

Münch, J.V. (1968) *Dokumente des geteilten Deutschland*, Stuttgart.

Ritter, G. and Hajdu, J. (1982) 'Die innerdeutsche Grenze', Cologne: *Geostudien*, 7.

—— (1989) 'The East–West German Boundary', *Geographical Review*, 79/3: 326–44.

Schwind, M. (1981) 'Kulturlandschaftliche Entwicklungen zu beiden Seiten der deutsch-deutschen Grenze', *Regio Basiliensis*, 22: 152–65.

Sharp, T. (1975) *The Wartime Alliance and the Zonal Division of Germany.*

Wagner, H. (1988) 'Die innerdeutschen Grenzen', in A. Demandt (ed.), *Deutschlands Grenzen in der Geschichte*, Munich.

Weigt, E. (1959) 'Standorte neuer Industriebetriebe in Franken und der Oberpfalz unter dem Gesichtspunkt von Nachbarschaft und Fühlungs-vorteil', *Berichte zur deutschen Landeskunde*, 23: 383–400.

Wulff, A. and Beyer, B.B. (1990) 'Die DDR und ihre Grenzgebiete zur Bundesrepublik Deutschland', *Raumforschung und Raumordnung*, 106–9.

3

THE NEW MAPS OF EUROPE
Fresh or old perspectives?
Michel Foucher

THE GEOPOLITICAL CONTRADICTION OF EUROPE

When considering the 'new' Europe and its crises, any analyst may wonder if it would not be more appropriate to consider 'old perspectives' – something like 'le retour du refoulé', according to the Freudian formula. Looking beyond the turmoil of the present, one could well ask just which, in reality, is nearer to the truth? Once again the question of European frontiers is back on the drawing-board and it was maps and boundaries that national leaders from the former Yugoslavia were discussing during the Geneva peace conference in January 1993, and it is about pieces of territory and front lines that armies are fighting in Bosnia.

It is possible to identify three stages in the development of the new European economic and political landscape. The first stage could be described as the period of economic and political division. Today the former Comecon (1949–91) is no longer an economic power, as its area of feeble economic activity represents no more than 1 or 2 per cent of world trade. The European Community has become the only pole of the economic continental area. Whereas the Community is indeed benefiting from the positive economic effects of a quarter of a century of industrial restructuring, everything remains to be done in the Eastern half in Europe, where the economic failures of past structures make current restructuring an arduous task. The other significant split is more institutional. On the one hand, the Western part of Europe is characterized by a high density of international organizations (one of the highest in the world: the European Community (EC), Council of Europe, Western European Union (WEU), North Atlantic Treaty Organization (NATO), European Free Trade Association (EFTA), Nordic Council, Baltic Conference, to name but a few); on the other, in the so-called

'Eastern Europe', the break-up of the Warsaw Pact (1955–91) and the Comecon has created a dangerous lack of supranational entities that could help to ensure both stability and development. The third split, essentially religious and historical, will be seen in greater detail below.

A second stage in this European recomposition could be described as the period of conflict. The disintegration of the Soviet Empire and the collapse of the people's democracies have promoted nationalisms and the nation-state concept. Crises have spread to many areas of the former Soviet Union, and numerous sub-regions of the 'evil Empire' are now independent and sovereign. These upheavals involve instability and conflict. Furthermore, Russia, the former core of the communist system, has to deal with many internal troubles. In the mean time three main areas are affected by ruptures in Central Europe: Romania, the Slovak and Czech republics, and the former Yugoslavia.

The third and more optimistic stage could be called the time of unity and agreement. Most of the Central European states are hoping to integrate into the European Community. Moreover, some of them are already planning to build regional structures of integration (for example, CEFTA between Poland, Hungary and the former Czechoslovakia).

Yet if the essential geopolitical contradiction in Europe is no longer the division of the continent by the 'Iron Curtain', it is an issue instead of Europe's multiple divisions. Europe numbers fewer states than nations or ethno-linguistic entities with a national vocation. This fundamental fact can be broken down into three questions along classic principles of geographical analysis. The first one, essentially philosophical, is about the Eastern limits of Europe. Does it follow the old 'Iron Curtain', the border zone between primarily Catholic and Orthodox religious zones, the western boundary of the former Soviet Union, or some other boundary further east? The second question will point out the intangible' nature of the state borders which, in fact, were defined by the Helsinki agreement under the concept of 'inviolability'. Must the principle of the right of states be ceded to the right of the people for self-determination? If so, what will the consequences be? Finally, what should the status be of internal boundaries of former federations that are disintegrating? To attempt an answer two issues have to be analysed, essentially from a geographical viewpoint, so as to determine which and where are the main 'fault lines' of the European continent and which is the best map of the present crisis, with particular attention to the case of the former Yugoslavia.

THE PRINCIPAL 'FAULT LINES' OF A WIDER EUROPE

Several lines or zones delimit the principal elements of a concept of a 'wider Europe' and serve equally to highlight the challenge of any geopolitical recomposition. One line is the old 'Iron Curtain' – the extreme limit of the Soviet Empire, defined at the outcome of the second European 'Civil War' (1939–45). This was an ideological front hardened by a strong military presence. Though diminishing now by peaceful negotiation, this is creating a strategic vacuum that the new regimes in Central Europe are not able to fill quickly. A new network of international institutions is required. This line shows the institutional limit which distinguishes existing EC, EFTA and NATO member-states from another Europe which has many years of economic restructuring before it. For example, the EC alone generates 80 per cent of the gross product on the European continent, not counting the Commonwealth of Independent States, which emphasizes the continued polarity of the 'wider Europe'.

The next line expresses the ancient division which distinguishes a Europe where the society was marked by Western Christianity and another Europe where Orthodox religion imposed its influence on the relationship between power and society and where it does not help to cast the foundations of a democratic society. It should be noted, however, that the emergence of the autocephalous church gave rise to distinct nations, notably Serbia, Bulgaria, Romania, and now the Ukraine and Russia.

From a geopolitical point of view, the most interesting division is the border area showing the old sociocultural frontier, to the east of which the efforts of modernization have, for centuries, followed a slower pace. It could be argued that communism was an effort to close the gap, but equally it accentuated the closed nature of the societies behind it. It does not always coincide with the western limit of the former Soviet Union, notably in the Baltic Republics and Western Ukraine. Interestingly, the Baltic states were among the first republics to reach independence while the second batch includes Lvov, the birthplace of Ukrainian nationalism, in this eastern part of the former Austrian Galicia. Further south, the line appears to cut the former Yugoslavia in half, though close examination reveals a much more complex situation. The line marks the demarcation of a two-speed progress towards modernization. The ideas of a 'return to Europe', democracy and the market economy are gaining ground quicker to the west. To the east, however, progress is slower and more haphazard. Nationalism is a possible strategy available to the communist parties and extremists.

If we observe the movement since 1989 towards a 'recomposition' of Europe, we can see that events have been focused within a Central Europe that is situated between the former 'Iron Curtain' and this eastern economic development boundary in Eastern Europe. So there is now a geopolitical choice in the conception of the 'wider Europe'. What is its eastern limit? Those who argue for a 'narrow Europe' suppose that the end of the settlement reached at Yalta can be accelerated by favouring the liberation of small nations, producing a set of alliances with strong reflections of the former Central European empires, notably the Austro-Hungarian empire. Let us take the case of Slovenia which is Catholic and populated by Slavs with a strong Austrian influence to their culture. An Austrian lawyer, Felix Ermacora, a specialist in international law, has maintained that preparation should be made for the 'reunification' of Slovenia with Austria (*Beitrittansuch*), a 'recuperation' which should begin with a customs union. As to Croatia, he wrote in *Die Welt* in July 1991 that 'it has always been Hungarian'. The old national or imperial solidarities are being used to justify a strategy of political recomposition. In France, for example, academic papers are increasingly appearing on the notion of Central Europe – *Mitteleuropa* – which substitutes for the idea of Eastern Europe, following the writings of Milan Kundera against the concept of Eastern Europe. Somehow, past historical structures seem more fundamental, more 'natural', than existing ones.

Another global sociocultural border worthy of consideration, however, is the one between the Orthodox world and the Muslim one. The war in Bosnia and the crisis in Kosovo have a clear relationship with this old featured border. The set of objective alliances between Serbia and Greece, for example, and Albania and Turkey is a modern version of an old story. This perception of a more general threat coming from the 'south' is growing in Russia, where Russian-speaking people are 'returning' from Central Asia. This second historical fault line seems to get a modern role in the new crises, from Sarajevo to Dushanbe and the Crimea.

THE MAP BACKGROUND: YALTA–POTSDAM AND VERSAILLES

Encouraging the idea of small nations in order to hasten the end of communism is a strategy that questions not only the territorial order of

Yalta, but also that of Versailles. It was the northern half of Central Europe that was modified by World War II (from 1939, date of the German–Soviet pact, until the 1945 treaty): Poland, the Baltic Republics, the western former USSR and Moldavia. These changes concerned roughly 30 per cent of the total length of the existing borders in Europe, some 7,700 kilometres (4,800 miles). The disputes originated in the Baltic Republics that had a tradition of statehood and sometimes of empire (Lithuania) as well as in Moldavia. In the case of Poland, the official policy is clearly to preserve the *status quo*. Although they are abandoning the political order of Yalta, they are not leaving the territorial order. The only case of change that has already taken place is the inter-German border. Of course, this was not a national frontier, but rather a 'front line', an ideological and military limit. It was, in addition, the only part of the 'Iron Curtain' which had created a political border.

In the southern part of Central Europe, however, most national boundaries date from the period 1910–20 and were simply re-established in 1945. Around 25 per cent of existing borders were drawn at the beginning of the century according to variable ethnic or strategic criteria. If this is the first difference with the northern part of Central Europe, there is a second one: the issue of ethnic minorities.

As a result of World War II, the percentage of minorities is now small in the northern part of Central Europe: 2 per cent in Poland, 3 per cent in Hungary and less than 5 per cent in Czechoslovakia (the Slovaks should not be considered a national minority but one of two nations of the former Federal Republic). In the southern half of Central Europe, however, the figures are much higher: 10 per cent in Albania, 11 per cent in Romania, 13 per cent in Bulgaria and 37 per cent in the former Yugoslavia (taking the sum of the non-native minorities and the 'Republican' minorities such as the Serbs located in Croatia).

A geographical examination of the minorities in this part of Europe shows that they are often located next to an old front line between the Austro-Hungarian and Ottoman Empires and within a zone of military borders (*Militärgrenze*). Each empire adopted a strategy of border control by a policy of agricultural colonization using 'soldier-peasants'. As the peasant population in these regions enjoyed religious and political privileges, there was a strong motivation to occupy what developed into frontier regions, causing an intermixing of different populations. In effect, the peasants served as front-line guardians. This in part explains the presence of a large concentration of Muslims in north-western Bosnia and Serbs in the *Krajina* ('border') region of Croatia, adjacent to Bosnia. The Croatian–Bosnian boundary follows very closely parts of the former Habsburg–Ottoman boundary established by the Treaty of Karlovci in 1699. These military borders extended from Croatia to the Moldavia-Bukovina region on the former-Soviet–Romanian border, along the River Sava north of Bosnia into Slavonia, in the Banat of Temesvar-Timisoara in Valachia, and into Transylvania. First introduced in the middle of the sixteenth century, they lasted within an obviously changeable zone limit, until 1881. These military borders were in a manner of speaking the 'Iron Curtain' of Western Christianity.

WAR AND FRONT LINES IN THE FORMER YUGOSLAVIA

The former Yugoslavia is a national and ethnic mosaic. The complexities of this recently formed country have been suddenly and viciously revealed. It was a country that was being shaken by a double crisis requiring profound adaptation: first, the exit of the communists, and second, not only the formation of new decision-making structures but also argument over the division of the political and financial competencies to the republics. Yugoslavia, born in 1918, was a geopolitical replica in miniature of the Soviet Union.

In its first phase it was a monarchical dictatorship that attempted to maintain cohesion of the kingdoms of the Serbs, Croats and Slovenians (the term 'Yugoslavia' was not used officially until after 1929). The civil war, linked to the Nazi occupation between 1941 and 1945, was followed by the installation of a self-governing authoritarian communism and the creation of three new republics, those of Macedonia, Bosnia-Herzegovina and Montenégro. A third phase developed after the death of Tito and the collapse of the communist bloc with which the country had remained closely tied as much on an ideological level as on one of trade.

During this third phase, during which the country tried to develop a new 'Yugoslav' formula, the crisis over the legitimacy of the communist apparatus aggravated the differences between the national groups. Nationalism here was not a primary factor heading ineluctably towards a 'collapse', but rather a tool available to the power-brokers and political opportunists in Yugoslavia. The encouragement of ethnic tension has favoured nationalist revival and burdened the political atmosphere at a moment when the Federal Constitution of 1974 has been judged outmoded, particularly in Slovenia and Croatia.

There is no lack of the necessary elements for such a strategy of tension and it is important to analyse them in a territorial perspective. The republican boundaries rarely follow ethnic boundaries (if such ethnic boundaries can ever be defined clearly). Certainly there is a core of founders that are reasonably ethnically homogenous (Slovenia is 90 per cent Slovenian, Croatia 75 per cent Croatian, Macedonia 67 per cent Macedonian and Serbia 66 per cent Serbian); but many regions are intermixed.

Within the hypothesis of partition, the challenge will first be to fix the limits of Croatia and Serbia. Croatian right-wing nationalists refer to 'Greater Croatia' – that of World War II – which includes all of Bosnia-Herzegovina and the area west of Vojvodina as far as the river Tisza. The Serb nationalists (since 1987, increasingly following the communist leader Slobodan Milosevic) are demanding the creation of a 'Greater Serbia' in case a single federation under Serb control is not installed. This includes all the regions, majority or minority, peopled by Serbs, which means Serbia, Bosnia (32 per cent Serbs to the west and south), East Croatia (11.5 per cent Serbs), Montenegro (of which the inhabitants are essentially of Serbian stock and who were much less affected by the Ottoman domination) and probably Macedonia (which the Serbs allege was drawn by Tito so as to weaken Serbia). Greater Serbia, according to the Movement of Serb Revival, will include 'all the regions and places where there are Serb cemeteries' or alternatively the 'states which formed part of the Kingdom of Serbia on December 1st, 1918, as well as the Regions which had Serb majorities before the pogroms between 1941 and 1945' – in other words, most of the neighbouring republics.

This break-up has carried enormous risks, particularly for the Muslims of Bosnia, who have witnessed the effective collapse of the Bosnian state and are under pressure primarily from Serbian, but also from Croatian, forces. Changes to the borders could have a significant impact beyond the former Yugoslavia's boundaries. Greece and Bulgaria have claims on the republic of Macedonia. There is growing concern in Albania for the Albanian population of Kosovo. Finally, the Hungarian government has hinted heavily that a new territorial settlement in Yugoslavia would re-open the question of Vojvodina, the autonomous Serbian province across the southern border, where almost half-a-million Hungarians live but where Serbs are in a majority (56 per cent versus 22 per cent). The Hungarians are located north of Novi Sad and form a local majority in eight districts in Vojvodina.

These views give us a key to understanding present pressures in favour of a more global reorganization of borders in Europe. They represent a sort of electoral geography – gerrymandering on a national scale.

The peace-plan proposals presented in Geneva in 1993 by Cyrus Vance and Lord Owen aimed to maintain a Bosnian State built on a constitution which would have recognized the legitimacy of the three main ethnic entities. The new Bosnia-Herzegovina would have become a decentralized state with 10 autonomous provinces (with separate legislatures, independent judiciaries and democratically elected administrative leaders): three provinces for each dominant ethnic group – Muslims, Serbs and Croats. Sarajevo would have become an 'open city' on the basis of a mixed community within a demilitarized area. The second peace-plan proposed by Thorwald Stoltenberg and Lord Owen aimed to reorganize the partition into a union of three ethnic republics: the Serbs would have received 52 per cent of the land, the Bosnian Muslims would have received 30 per cent and the Croats 17.5 per cent. During the year 1993, all these groups appeared to be waiting to see what the reaction in Western Europe and the United States would be to the Vance–Owen and Owen–Stoltenberg plans; ultimately the conflicting parties will dictate whether the war ends or whether it will spread over other regions (particularly Kosovo and Macedonia). It remains to be seen whether ethnic cleansing will be effective in 'gerrymandering' the ethnic geography of Bosnia.

It would appear, then, that the time has come for a general review of borderlands and boundaries in Europe, including those of the beginning of the twentieth century, in the hope that this can be done by peaceful means. There is a lot of romanticism behind such ideas. However, religious and ideological rivalries are some of the most powerful border-making factors. It is as if Europe cannot create boundaries without an adversary. It used to be the Barbarian, then the Turk. Only yesterday it was the Soviet Union. What about tomorrow . . .?

4

PEACEKEEPING LESSONS FROM DIVIDED CYPRUS

Carl Grundy-Warr

INTRODUCTION

The focus of this chapter is on the role of the peacekeepers of the United Nations (UN) in the intercommunal conflict between the Greek and Turkish Cypriots. It examines how the island's *de facto* political geography has been completely transformed since Cyprus's independence from British colonial rule in August 1960. The analysis concentrates on the ground-level activities of the UN blue berets,[1] rather than on the higher-level diplomatic peace-making efforts of the UN to get the disputants to negotiate a political settlement.[2] The UN Force in Cyprus (UNFICYP) has had to interact with ordinary Cypriots directly affected by political division on a daily basis for almost thirty years. Thus, the peacekeepers have become features of the peculiar political landscape of the island.[3] A better understanding of their role helps to provide an insight into the limitations and potentialities of third-party multinational intervention in conflicts involving the political partition of territory.

At the time of writing, the size of UNFICYP has been cut to a level that may undermine its operational effectiveness. The cuts have come at a time when the demands on UN peacekeeping are enormous. Since 1988 the UN has mounted more peacekeeping ventures than in the entire post-World War II period. New missions have been sent to Angola, El Salvador, Western Sahara, Croatia and Bosnia-Herzegovina, Somalia, Cambodia, and the Iraq–Kuwait borderland. The range of tasks these forces have been given ranges from inter-state border monitoring to complex intra-state roles, including civil administration; the supervision of elections; civil law and order; humanitarian relief to civilians affected by ongoing warfare; rehabilitation of displaced

71

persons; de-mining programmes; and infrastructural projects for war-damaged societies.[4]

In short, the UN is enjoying a much broader role in the management of conflicts in the post-Cold War era, and peacekeeping must be seen as a multi-dimensional activity. Indeed, the current UN Secretary-General, Boutros Boutros-Ghali, has encouraged 'a more ambitious approach to conflict management' (Boutros-Ghali 1993: 67). This involves closer co-ordination between three key dimensions of the conflict management process – peacekeeping, peace-making and peace-building, which involves several of the broader roles that have been assigned to recent UN operations (see above) that are aimed at trying to create on-the-ground conditions for a peaceful settlement.

A key argument of this chapter is that, whilst nobody can deny a much greater role for UN peacekeeping in the 1990s than in any previous decade, there are many lessons still to be learnt by looking at one of the longest-standing UN missions. In Cyprus the UN have played two entirely different roles under the same mandate: between 1964 and 1974 UN blue berets effectively played the part of an intercommunal constabulary force, but in the period since *de facto* partition in 1974 to the present time UNFICYP has been a buffer force. Whilst the underlying principles of the peacekeeping mission in Cyprus have remained constant, the dramatic changes in the political landscape of the island entailed considerable adjustment and change to the blue berets' day-to-day activities. The experience of UNFICYP in dealing with totally different political and military situations should help to inform current and future UN operations elsewhere dealing with either intra-state or inter-state conflicts.

Some writers have argued that the peacekeepers in Cyprus share responsibility for the maintenance of an unsatisfactory political status quo. For instance, Coufoudakis (1976: 459) argued that during 1964–74 UNFICYP 'also froze the situation so that the Turkish Cypriot enclaves and lines of division in the cities remained intact for a decade'. More recently, Mandell (1992: 221) has argued that UNFICYP has been 'among the more important factors contributing to the continued stalemate of the Cyprus problem' and that the Force has helped to 'institutionalize the conflict' and reduce the motivation for serious negotiation between the political leaders of the two main communities of Cyprus. In fact, this author's research shows that the maintenance of the cease-fire (and the existing *de facto* partition) is an essential prerequisite to any meaningful political negotiations over the future of the island. Furthermore, UNFICYP commanders have always sought to encourage efforts to bridge the divisions that exist between the Greek and Turkish Cypriots. They have done this by means of ground-level essential humanitarian activities. A large part of the following discussion is an empirical analysis of these important aspects of UNFICYP's role.

GREEN LINES AND BLUE BERETS

At independence in August 1960, Cyprus had a population of 650,000, out of which Greek Cypriots outnumbered Turkish Cypriots in a ratio of 4 : 1. The island had 114 mixed villages, and in spite of distinct ethnic quarters in the main towns and in several villages, the two communities co-existed freely in many villages. Unfortunately, the Independence Constitution, which was the product of the London–Zurich Agreements of 1959 between three outside powers – Britain, Turkey and Greece – polarized community politics. After the collapse of the constitution, intercommunal violence flared up in Nicosia and spread into the countryside and district towns. Between December 1963 and August 1964 the political geography of the island was completely transformed. The Government of Cyprus (now a Greek Cypriot administration, due to the withdrawal of Turkish Cypriot officials) could not prevent the Turkish Cypriot community from making the *de facto* control of territory a key political issue. Turkish Cypriots had evacuated 72 mixed villages and 24 wholly Turkish Cypriot villages. United Nations figures estimated that there were approximately 25,000 refugees, of whom 21,000 were given homes in larger or 'safer' Turkish Cypriot settlements, while the remainder found temporary shelter in refugee camps. (UN document S/6102, 12 December 1964, para. 45). Approximately 55,000 Turkish Cypriots relied mainly on food and medical supplies shipped from Turkey (Crawshaw 1978: 368).

Between December 1963 and March 1964 the British Army acted as peacekeepers. This was largely because they already had a presence on the island in the two British Sovereign Base Areas of Dhekelia and Akrotiri, and Britain was one of the Guarantors of the Independence Agreement. Even so, the British Army was clearly remembered by the Greek Cypriots for its part in trying to suppress the EOKA (*Ethniki Organosis Kyprion Agoniston*, National Organization of Cypriot Fighters) guerrilla campaign against colonial rule and in favour of *Enosis* (Union with Greece) during the so-called 'Emergency Period' from 1955 to 1959. The leaders of the anti-Turk violence in the 1960s were mostly drawn from the ranks of former EOKA fighters. Despite the

73

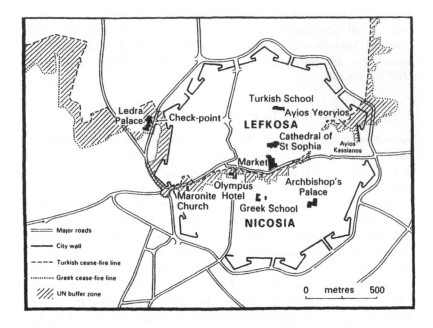

Figure 4.1 Divided Nicosia: this map shows the UN buffer zone running through the heart of the capital of Cyprus separating the Greek Cypriot and Turkish Cypriot quarters of the city. In places the buffer zone is barely 10 m across, and it is once again being patrolled by British troops, following the departure of the Canadian contingent in June 1993. Whilst this ethnic division was reinforced by the Turkish military partition of the whole island in July–August 1974, in fact the line follows almost the same path as the old 'green line' drawn by a British Army officer on 28–29 December 1963, following intercommunal fighting in the city. Since that date, the Green Line has become a lasting symbol of the Cyprus Problem, and a line of international geopolitical significance

Source: UNFICYP, but map drawn by Carl Grundy-Warr.

ex-colonial army's awkward role as post-colonial peacekeepers they were involved in preparing the ground for the UN. One of their first tasks was to segregate the communities in the zones where inter-communal violence was most acute.

The most important division was in the heart of the old walled city of Nicosia, and it roughly corresponds with the present-day partition (Figure 4.1). It was on the night of 28–29 December 1963 that a green chinagraph pencil line was drawn across a map of the city by Major-General Peter Young under the auspices of representatives of both communities. The Green Line was never intended to be anything more than a temporary *cordon sanitaire* between the quarters of the two main communities of Cyprus. In fact, it solidified into a lasting symbol of ethnic segregation and 'an unremitting obstacle to progress towards normalization between the two communities' (Harbottle 1970: 67). As Stagenga (1968: 40) noted, the creation of this and lesser-known 'green lines' in Larnaca and other towns, meant that the British were 'tacitly partitioning parts of the country' prior to the arrival of the blue berets in March 1964. And as soon as UNFICYP was authorized

> Both Cypriot communities were aware that once this force was deployed the then existing patterns of coercive control throughout the island would be 'frozen'. Both sides therefore were intent on consolidation or extending their control before UNFICYP could intervene.
>
> (Patrick 1972: 60)

In February 1964 President Makarios, the Greek Cypriot head of the Republic of Cyprus, rejected a United States proposal for an enlarged peace force made up exclusively of NATO contingents to Cyprus. The Greek Cypriot leadership was mindful of the strong left wing, led by AKEL (Reform Party of the Working People), who were against a further extension of Western influence in Cyprus. Furthermore, Makarios occupied a position of leadership in the non-aligned movement, which ruled out Cyprus as a strategic NATO base. In addition, the Greek Cypriot leadership believed that their aims could best be realized through the UN. On 4 March 1964 the Security Council unanimously adopted Resolution 186 authorizing, with the consent of the Government of Cyprus (minus the Turkish Cypriot leadership), the formation of UNFICYP.

At its peak strength, in June 1964, UNFICYP comprised 6,400 personnel.[5] They deployed blue berets at all the key trouble-spots and ethnic interfaces where confrontation was most likely. They arranged local cease-fires, manned and demarcated various 'neutral zones' and negotiated for the removal of fortifications, road-blocks and other physical barriers wherever they could. However, by March 1964, Cyprus was already a republic 'riddled with holes' (Drury 1981: 298) or tiny pockets of Turkish Cypriot control under the protection of their own paramilitaries.

INTERPRETATION OF THE UN MANDATE

The UN Security Council Resolution 186 recommended

> that the function of the Force should be in the interest of preserving international peace and security, to use its best efforts to prevent a recurrence of fighting and, as necessary, to contribute to the maintenance and restoration of law and order and a return to normal conditions.

Such an ambiguous mandate left plenty of room for differing interpretations of UNFICYP's responsibilities in Cyprus. In practice, the UN Secretary-General is delegated substantial authority by the Security Council to clarify and interpret the UN mandate for the practical purposes of a peacekeeping operation. U Thant's interpretation of the mandate conflicted with the diametrically opposed perspectives of the Greek Cypriot and Turkish Cypriot leaderships, who saw UNFICYP's tasks in terms of their own mutually exclusive political goals. The Greek Cypriot leadership wanted to extend their effective territorial control over the whole Republic, including the forty or so areas controlled by the Turkish Cypriots. As the internationally recognized administration for the whole island,[6] it argued that the UN Force should help it to eliminate the Turkish Cypriots' 'rebellion' in order that peace, law and order could be restored. As the Greek Cypriot Interior Minister, Georgadjis, put it,

> The United Nations troops should be neutral, but that does not mean they should treat both sides on an equal footing in the conflict ... You cannot equate a majority with a minority, the legal government with the leaders of a group.
>
> (*Cyprus Mail* 1964: 1)

References were frequently made by Greek Cypriot leaders to the part played by the UN Operation in the Congo (Opération des Nations Unies au Congo, or ONUC) in crushing the Katanga secession, as if to imply that UNFICYP should play a similar role in Cyprus. In one sense, UNFICYP conformed to the Cyprus Government's conception of its duties in as far as they were trying to remove the military lines of division that had developed between the two communities. Nevertheless, U Thant was insistent that UNFICYP should not be perceived as an adjunct of the Makarios Administration.

Indeed, UNFICYP's pacification efforts between 1964 and 1968 meant that blue berets frequently had to keep the protagonists apart.

Some Greek Cypriot leaders criticized such efforts as a deliberate attempt to protect the Turkish Cypriots in their 'separatist' cause. Even so, UN plans to dismantle physical barriers put up by both sides put the Turkish Cypriots at risk, because UNFICYP could not guarantee their safety from the much stronger and better-armed Greek Cypriot militias. The Turkish Cypriots[7] saw UNFICYP's primary role as one of protecting their community, split up into several enclaves throughout the island, from attacks by the Greeks. This the UN soldiers on the ground tried to do, although not with the purpose of preserving a patchwork territorial division of Cyprus, but simply with the intention of saving lives and preventing an escalation of the internal conflict.

According to one observer,

> the mandate of UNFICYP was vague enough to allow the disputants to read their own self-serving interpretations into it ... Though this ambiguity in UNFICYP's mandate could and did lead to problems in the field, it helped to make the UN Force acceptable to the concerned parties.
>
> (Moskos 1976: 37)

As with other UN peacekeeping missions, it was impossible to please both parties at once. Whilst the UN continued to recognize the authority of the Government of Cyprus without the Turkish Cypriot leadership, UNFICYP was determined not to be seen 'as an arm of the government' (U Thant, UN Doc. S/5950, 10 September 1964, para. 220). U Thant realized that in the absence of a political settlement the best UNFICYP could do was to maintain an 'uneasy equilibrium' (U Thant, UN Doc. S/6228, 11 March 1965, para. 274). With regard to the highly ambiguous part of the mandate asking the Force to contribute to 'a return to normal conditions', UNFICYP has always avoided any particular political interpretation of this phrase. Rather, the UN peacekeepers sought to render assistance 'in the amelioration of day-to-day administrative, economic, social or judicial difficulties arising from the division of the communities' (U Thant, UN Doc. S/6102, 12 December 1964, para. 24).

With respect to the prevention of a recurrence of fighting in the island, UNFICYP did manage to succeed in reducing the monthly average of armed engagements from 3,000–4,000 to less than 10 by 1968.[8] Nevertheless, UNFICYP's success in managing an uneasy ceasefire throughout the island was not matched by any movement towards a political settlement. Some observers of the Cyprus scene have criticized UNFICYP's role for helping to preserve this negative stability

(Coufoudakis 1976). In fact, the blue berets did much to ease local tensions between the two Cypriot communities. It was the actions of the respective community leaderships, their outside backers and the armed militias within Cyprus, particularly on the Greek Cypriot side, that tended to harden the divisions on the ground. On the one hand, the Turkish Cypriot Administration refused to co-operate with the Greek Cypriots in any project that involved an extension or recognition of the Cyprus Government's authority in the territory under the *de facto* control of the Turks. Sowerwine (1992: 6) has argued that mainland Turkey encouraged the Turkish Cypriot enclaves and self-segregation 'in the belief that the longer a separate Turkish Cypriot administration functioned on Cyprus, the better the odds of securing a geographical separation for the Turkish community'. On the other hand, the Cyprus Government only reinforced the ethnic segregation by building new infrastructure around the Turkish Cypriot enclaves, which deprived the Turkish Cypriots of basic services and amenities. It is true that UNFICYP's humanitarian aid to the Turkish Cypriots helped to maintain enclaves, but to ignore their plight would only have helped the cause of the anti-Turkish elements amongst the Greek Cypriots. Without the presence of the blue berets there would undoubtedly have been greater bloodshed, and it is likely that Turkey would have intervened in the conflict sooner than it eventually did.[9]

UNFICYP cannot be blamed for the failure to reach a political compromise during the intercommunal talks after 1968. In fact, an UNFICYP proposal that the military operation should give way to a civilian counterpart in which all the UN specialized agencies should be involved, working together as a single entity to help rehabilitate and reconstruct the shattered relations between the two communities, was not accepted by the Security Council.[10] UNFICYP's efforts to restore a measure of mutual trust between divided communities helped to create a period of relative calm, but it was the failure of the peace-making process that froze the existing patchwork partition of the island.

THE UN BUFFER ZONE

The summer of 1974 will live long in the memories of many Cypriots. How can they forget? There is a permanent reminder of the losses, the bloodshed and atrocities, and of years of intercommunal strife, in a scar etched across the landscape of the island, separating Greek from Turkish Cypriots. The physical division of the island on an ethnic basis turned about one-third of the population of Cyprus into refugees and involved enormous changes to the human geography of the island. The events leading up to the Greek *coup d'état* against Makarios and the subsequent Turkish military intervention have been discussed at length elsewhere.[11] What is of interest here is how UNFICYP's role was changed by the creation of an effective *de facto* border between the two main Cypriot communities.

UNFICYP had neither the authority nor the ability to prevent the Turkish Army from occupying 37 per cent of Cyprus. Blue berets did what they could to assist refugees, arrange local cease-fires and protect civilians living in vulnerable enclaves and villages. The UN Force also held on to Nicosia International Airport which the Turks tried to capture. Since then the airport has not been used, within a UN-protected area.

When an island-wide cease-fire was declared at 1600 hours on 16 August 1974 the political map of Cyprus had been arbitrarily and violently rearranged. Suddenly the political situation was changed

> from an argument over how the intermingling of two different populations was to be regulated – by minority rights, by power-sharing, or by devolution – into a different kind of argument, one about what sort of federal link could be built between territorially separate communities.
>
> (Kyle 1984: 15)

An artificial line cut through the island like a cheese-wire, cutting off villages from their fields, splitting streams and underground water resources, truncating roads and power-lines. One of the immediate tasks of the UN peacekeepers was to demarcate a demilitarized area between the Turkish Army cease-fire line and the forward positions of the Greek Cypriot National Guard. The eventual buffer zone was 180 kilometres long, from the isolated Kokkina enclave and Kato Pyrgos in the north-west to the south-east coast south of Famagusta (Figure 4.2). Between July 1974 and December 1975 some 185,000 Greek Cypriots moved south and about 45,000 Turkish Cypriots moved northwards. In February 1975, the northern *de facto* state was named the 'Turkish Federated State of Cyprus',[12] which was totally dependent on mainland Turkey for economic support and security.

The key tasks of UNFICYP in the eighteen months following *de facto* partition can be summarized as the maintenance of the cease-fire, the interpositioning of troops between the forward positions of the opposing armies, the protection of the Greek Cypriots and Turkish Cypriots still living on the 'wrong' side of 'the Attila Line' (the name

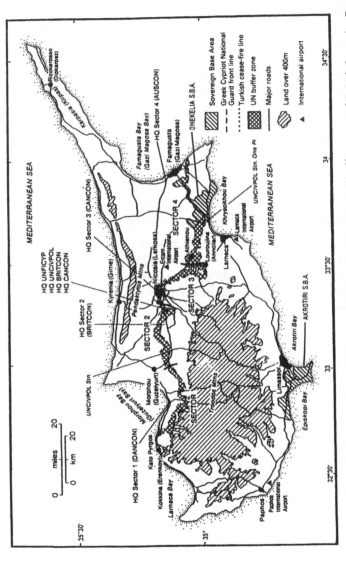

Figure 4.2 The *de facto* partition of Cyprus: map of the island of Cyprus detailing the UN buffer zone and British Sovereign Base Areas. The map shows the operational sectors of the contingents of UNFICYP before the 1992–3 reductions in the Force. The withdrawal of the Canadian and Danish troops left only the British and Austrians in charge of monitoring the buffer zone prior to the arrival of a small Argentinian contingent in October 1993. In addition to these military contingents there is a small number of UN civilian police from Australia and Sweden (since replaced by Irish police) whose job it is to liaise between the respective police authorities on both sides of the buffer zone, and to deal with law and order problems within the UN-monitored zone. There are inhabited villages within the demilitarized area, such as Athienou in Sector Four, and also Pyla, which is effectively the only village with both Greek and Turkish Cypriots still living in the same place.

Source: UNFICYP. Drawn by the Cartographic Unit, University of Durham.

given by the Greeks to the partition) and the support of relief operations co-ordinated by the United Nations High Commission for Refugees (UNHCR) and the International Committee of the Red Cross (ICRC).

During the last nineteen years UNFICYP has kept cease-fire violations to a minimum and it has maintained the unsatisfactory military status quo, as required by the unchanged mandate. On many occasions the UN has had to deal with minor adjustments to front-line positions, encroachments into the demilitarized area and attempts by Greek Cypriot civilians to cross over the buffer zone. For many a lonely sentry stationed at isolated observation posts in the Cypriot country-side, the tour of duty with the UNFICYP has been a very tedious affair. Even the movement of a sandbag or a barrel by half a metre into the buffer zone is enough to involve young UN officers in delicate negoti-ations with the offending soldiers and commanders on either side to restore the status quo. The mundane nature of many of UNFICYP's daily tasks and level of vigilance expected was summed up by the words of one former UNFICYP officer, 'When a frog croaks in the buffer zone the (UNFICYP) commander wants to know about it.'[13] It is the success of the purely military dimension of the peacekeepers' role on Cyprus that has facilitated efforts to lessen the disruption to everyday life created by the division of the island's territory.

SOCIO-ECONOMIC AND HUMANITARIAN ACTIVITIES

Since partition, UNFICYP has had less opportunity than in the pre-1974 period to encourage local-level 'peace-building' through the initi-ation of bi-communal projects, or, at least, occasional meetings between Greek and Turkish Cypriot villagers to solve local resource management problems. Even so, the UN buffer zone contains about 3 per cent of the island's territory, including valuable agricultural land, water resources, inhabited (mostly Greek Cypriot) villages and the Green Line of Nicosia. Without the presence of UNFICYP all this zone could have become a highly dangerous area of close confrontation or a no-man's-land for civilian activities. Slowly since 1974, UNFICYP have helped to extend economic activities, particularly farming, to within metres of the Turkish Army front-line positions. In fact, the buffer zone is one of the very few remaining areas of Cyprus where intercommunal activity and a limited degree of co-existence is possible (Grundy-Warr 1987 and 1993). UNFICYP have also assisted both communities on matters relating to the cross-buffer zone supply of electricity and water, and it has arranged for meetings between divided families at the Ledra Palace Hotel. Blue berets have also been responsible for providing certain humanitarian services to the isolated and diminishing number of Greek Cypriot inhabitants (about 800) on the Karpas peninsula on the Turkish side, and the even smaller number of Turkish Cypriots still living in south Cyprus.

All the socio-economic activities of UNFICYP involve UN personnel in day-to-day contacts with ordinary Cypriots, *mukhtars* (village leaders), farmers, police officers, army officers, bureaucrats and politicians from both communities. They are aspects of the ground-level, hands-on micro-diplomacy that has become essential for the success of UN peacekeeping ventures, especially in locations where the UN has been established for long periods, such as in Cyprus and in Southern Lebanon. Such activities have helped to preserve at least a measure of mutual trust between ordinary Cypriots whose daily lives are most adversely affected by the existence of a military-imposed partition.

REDUCTIONS IN UNFICYP'S SIZE

It would be unfortunate, to say the least, if the present effectiveness of UNFICYP should become the pretext for failure to find a solution to the fundamental problem of Cyprus.
(U Thant, UN Secretary-General, UN Doc. S/6102, 12 December 1964, para. 239)

Over twenty-eight years after U Thant wrote these words Boutros-Ghali is having to deal with a rapid reduction in the size of UNFICYP due to the frustration of the troop-contributing states with the continuing political deadlock in Cyprus and the dire condition of UN finances. After the failure of the 1992 round of intercommunal talks under UN auspices, the political impasse continues and the prospects for a long-term settlement to the Cyprus Problem seem as distant as ever.

Before the UN-sponsored talks in 1992, Boutros-Ghali highlighted the problems of financing UNFICYP and the growing dissatisfaction of the troop-contributing governments with the lack of progress in Cyprus.

If a Force has for 28 years maintained conditions in which a peaceful settlement to a dispute can be negotiated but negotiations have not succeeded, it has to be asked whether that Force has a priority claim on the scarce resources which Member States can make available to the Organization's peace-keeping activities.
(UN Doc. S/23780, 3 April 1992, para. 33)

Since then the Force has undergone big reductions with the departure of two of the longest-standing contingents in peacekeeping history. The Danish contingent (DANCON) left the island in December 1992. Over the years the Danes have contributed 57 contingents, more than 30,000 soldiers, and large amounts of goodwill. Early last year the Danish government decided that the Cyprus battalion 'was more needed else-where' (*Cyprus Weekly* 1992: 21). Denmark also had troops serving with the UN in Croatia, Bosnia-Herzegovina and Kuwait. In addition, UNFICYP has now lost the Canadian contingent (CANCON) after twenty-nine years' service. The Canadian pull-out started in June 1993, and it has effectively reduced the strength of the Force to below 1,000 personnel. Before the changes the 180-kilometre-long buffer zone was divided into four key sectors, controlled from west to east by the Danes, the British (BRITCON), the Canadians and the Austrians (AUSCON) respectively (see Figure 4.2).

Troop withdrawals have forced UNFICYP to restructure its oper-ations. At the time of writing, the Force has around 1,400 personnel, which includes 1,050 peacekeepers drawn from Britain, Austria, and Argentina.[14] UNFICYP also incorporates 30 civilian police from Australia and Ireland, and a 12-man observer group made up of army officers from Hungary, Austria and Ireland, who are an integral part of the mission. Reductions in strength has led to more mobile patrols and fewer permanently manned observation posts. The constraints imposed by a much-reduced physical presence may well hamper UNFICYP's ability to respond quickly to cease-fire violations and inci-dents in or near to the demilitarized zone.

More than ever, UNFICYP is reliant upon the restraint and coopera-tion of parties on both sides of the buffer zone to avoid local-level conflicts from arising. An example should help to illustrate this point. Villagers from Athienou, close to the cease-fire line, held a small demon-stration in the buffer zone. They wanted to cross the *de facto* partition in order to visit the chapel of Saint Epipharios located in the Turkish-held north. About 300 men, women and children were carrying Cypriot flags and placards declaring 'End the occupation of our lands'. One kilometre away, behind the Turkish sentry-points, were a group of Turkish Cypriots putting on a counter-demonstration. Between them were a dozen or so blue berets, who would have been powerless to prevent the Greek Cypriot demonstrators from marching towards the Turks had they continued to do so. Fortunately they were content to hand over a formal written protest to the UN officer on the spot (*Cyprus Weekly* 1992: 20). Peaceful demonstrations like this could lead to localized violence that would create serious intercommunal tensions if allowed to cross the buffer zone. In April 1993 a young Greek Cypriot National Guardsman was killed in a shooting incident along the Green Line of Nicosia. Every such incident raises the political temperature and damages the efforts of those who are trying to develop mutual trust between the two communities.

CONCLUSIONS

Peacekeeping operations do not operate in a political vacuum, and their activities can only be understood in context. As one astute observer of international peacekeeping ventures has stressed, peacekeeping 'is dependent, in respect of both its origins and its success, on the wishes and policies of others' (James 1990: 5). Those who argue that UNFICYP's presence has reduced the urgency to search for a solution to the Cyprus Problem miss some essential points about the nature of peacekeeping. As Professor Alan James stated in an interview in 1989:

> If UNFICYP is withdrawn in order, as some people suggest, to 'bring the Cypriots to their senses', the urgency of finding a solution in Cyprus may increase, but this does not mean that the two sides would come closer to resolving their differences, given their perception of their respective rights and security interests.[14]

The UN has had success in the maintenance of the military status quo in Cyprus, but this has mostly been due to the preference of both sides for peaceful ways to settle their differences. Some 30,000-plus Turkish troops in the north of Cyprus could not be prevented from invading the south by a thin blue line of UN soldiers. The fact is that peacekeeping forces are not capable of preventing large-scale military confrontations. The valuable role UNFICYP plays is in acting quickly to prevent minor incidents and cease-fire violations from escalating into major ones. In the course of their day-to-day duties along the buffer zone, the blue berets practise a form of micro-level diplomacy, without which it is unlikely that higher-level peace-making would be possible.

FUTURE ROLES

Many dimensions of the so-called Cyprus Problem involve an inter-twining of the past with the present. Physical reminders of painful past conflicts exist all over Cyprus. This is especially so in the empty streets and abandoned buildings in the 'dead zone' along the Green Line of central Nicosia. The UN blue berets have become important elements of the partitioned landscape of Cyprus, but they are not responsible for the preservation of the unsatisfactory political division. Indeed, they have helped to reduce cross-buffer zone tensions, although this is a far cry from anything resembling 'normal conditions'. Any future steps towards a settlement will probably involve UNFICYP in peacekeeping and limited 'peace-building' activities.

Many of the 'confidence-building measures' highlighted in UN Resolution 789 (UN Doc S/24841, 24 November 1992) imply an enlarged role for UNFICYP, which seem at odds with the reductions in the size of the operation. Amongst the measures suggested are the further de-manning of positions on either side of the buffer zone; the implementation of UN Resolution 550 (1984) calling for the hand-over of Varosha, a former Greek Cypriot suburb of Famagusta, to UNFICYP; the promotion of people-to-people contacts by reducing restrictions of movement across the buffer zone; ending restrictions on the cross-border movements of foreigners; and the implementation of bi-communal projects. Such steps would certainly soften the almost impenetrable character of the current *de facto* partition separating the two sides, and they could not take place without an international third party to help manage the changes.

At the time of writing, the UN Secretary-General faces problems trying to persuade UN member-states to contribute troops and funds to the long-serving mission at a time when there are more pressing demands for peacekeepers elsewhere in the world.

Although troop-contributing countries meet 70 per cent of the cost of UNFICYP, the UN Force has an accumulated debt of nearly $200 million, because it relies on voluntary contributions to make up the remaining 30 per cent. Contributions have never been sufficient to meet the cost of the Force and troop-contributing countries have not been reimbursed for their expenditures since December 1981 (Security Council / 5519, 14 December 1992). On 13 May 1993 a British-sponsored Security Council resolution to convert the funding of UNFICYP from voluntary to assessed contributions was blocked by a Russian veto. Whilst the Greek and Cyprus Governments have offered annual costs of UNFICYP, there must be some question marks over the longer-term viability of the operation.

Just as UNFICYP faces a leaner and a possibly more testing future, the fate of a Cyprus settlement now rests with the fourth president of the Cyprus republic, Glafkos Clerides, and his one-time lawyer colleague and fellow interlocutor in the intercommunal talks between

1968 and 1976, the Turkish Cypriot leader, Rauf Denktaş. In an island where so many aspects of the past weigh heavily on the present it is hoped that this old political formula will break the uneasy deadlock in negotiations. If not, Cyprus will continue to have a heavily monitored *de facto* border preventing intercommunal contact between Cypriots for many years to come.

NOTES

1 Peacekeepers in Cyprus wear blue berets.
2 For a discussion of the theory and practice of peace-making refer to Miall 1992.
3 For an elaboration of the political geography of peacekeeping in Cyprus refer to Grundy-Warr 1993, and for a discussion of 'peacekeeping landscapes' refer to Brunn 1991.
4 Professor Alan James has argued that we should be careful not to consider all the new demands placed on peacekeeping operations simply as post-Cold War phenomena. Several of the Cold War UN operations have had to deal with complex 'internal' situations. These include the Congo mission (ONUC), the Cyprus mission in its first decade of operations, the Dominican Republic operation in the mid-1960s (Domrep), and in practice, the UN Interim Force in the Lebanon (UNIFIL), to name but a few. An excellent source on both UN and non-UN peacekeeping operations is James 1990.
5 Initially, the main contributing countries to UNFICYP were Britain, Canada, Sweden, Ireland, Finland, Austria and Denmark.
6 As early as 27 December 1963, President Makarios informed the UN Secretary-General that only the Permanent Representative of Cyprus, Zenon Rossides, was authorized to speak for the Government of Cyprus. The fact that the UN Security Council implicitly accepted this, and have never recognized any of the Turkish Cypriot administrations as legitimate, remains in effect thirty years later.
7 For an interesting discussion of the Turkish and Turkish Cypriot perspectives of the Cyprus conflict, see Sowerwine 1992.
8 The major acts of aggression during the years 1964–8 were by the Greek Cypriot fighters against Turkish Cypriot settlements, particularly Kokkina, Kophinou and Ayios Theodhoros. For details see Harbottle 1970, Stagenga 1968 and Patrick 1976.
9 The attack on Kophinou in 1967 nearly provoked a Turkish military retaliation, which was prevented only by the exercise of shuttle diplomacy by Cyrus Vance (later President Carter's Secretary of State). Even so, not all public opinion in Turkey was placated by diplomatic action, and there was strong pressure on Ankara to resort to arms to protect the beleaguered Turkish Cypriot enclaves in Cyprus (see Sowerwine 1992).
10 Brigadier Michael Harbottle, who was Chief of Staff of UNFICYP from June 1966 to August 1968, has long argued that one of the biggest failures of the Security Council towards Cyprus was not broadening the mandate

for the UN operation to incorporate more peace-building measures between the communities at a time when intercommunal fighting had virtually ended (Harbottle 1970, and *pers. comm.*). As a result, UNFICYP's role remained a mostly static military one of maintaining the island-wide cease-fire up to the tragic events of the summer of 1974.

11 See Loizos 1976, Kyle 1984, Oberling 1982 and Hitchens 1989.
12 On 14 November 1983 the Turkish Cypriot leadership re-named their northern *de facto* state the 'Turkish Republic of Northern Cyprus'.
13 Interview with the author.
14 Alan James, interviewed in Stavrinides 1989.

REFERENCES

Boutros-Ghali, B. (1993) 'UN Peacekeeping in a New Era: A New Chance for Peace', *The World Today*, 49/4: 66–70.
Brunn, S.D.C. (1991) 'Peacekeeping Missions and Landscapes', in D. Rumley and J.V. Minghi (eds), *The Geography of Border Landscapes*, London: Routledge, 269–98.
Coufoudakis, V. (1976) 'United Nations Peacekeeping and Peacemaking and the Cyprus Question', *Western Political Quarterly* (USA), 29/3.
Crawshaw, N. (1978) *The Cyprus Revolt: An Account of the Struggle for Union with Greece*, London: George Allen & Unwin.
Cyprus Mail (1964), Nicosia, 20 March.
The Cyprus Weekly (1992), Nicosia, 18–23 December.
Drury, M.P. (1981) 'The Political Geography of Cyprus', in J.I. Clarke and H. Bowen-Jones (eds), *Change and Development in the Middle East*, London: Methuen, 289–305.
Grundy-Warr, C. (1987) 'Political Division and Peacekeeping in Cyprus', in G.H. Blake and R.N. Scholfield (eds), *Boundaries and State Territory in the Middle East and North Africa*, Cambridge: Menas, 70–84.
—— (1993) 'A Comparative Political Geography of United Nations Peacekeeping: Cambodia and Cyprus', *GeoJournal*, October (forthcoming).
Harbottle, M. (1970) *The Impartial Soldier*, London: Oxford University Press.
Hitchens, C. (1989) *Hostage to History: Cyprus from the Ottomans to Kissinger*, New York: The Noonday Press.
James, A. (1990) *Peacekeeping in International Politics*, Basingstoke: Macmillan in association with the International Institute for Strategic Studies.
Kyle, K. (1984) *Cyprus*, London: Minority Rights Group, Report No. 30 (second edition).
Loizos, P. (1976) *Cyprus*, London: Minority Rights Group, Report No. 30 (first edition).
Mandell, B. (1992) 'The Cyprus Conflict: Explaining Resistance to Resolution', in N. Salem (ed.), *Cyprus: A Regional Conflict and its Resolution*, New York: St. Martin's Press Inc., 201–7.
Miall, M. (1992) *The Peacemakers*, Basingstoke: Macmillan.
Moskos, C.C. (1976) *Peace Soldiers: The Sociology of a United Nations Military Force*, Chicago: University of Chicago Press.

Oberling, P. (1982) *The Road to Bellapais*, New York: Columbia University Press.

Patrick, R.A. (1976) *Political Geography and the Cyprus Conflict, 1963–1971*, Ontario: University of Waterloo Press.

Sowerwine, J. (1992) 'Conflict in Cyprus: The Turkish Dimension', paper presented at the 'Sixth Annual Conflict Studies Conference', Fredericton, N.B., Canada.

Stagenga, J.A. (1968) *The United Nations Force in Cyprus*, Columbus, Ohio: Ohio State University Press.

Stavrinides, Z. (1989) 'UNFICYP: Preserving the Stalemate', *The Cyprus Weekly*, Nicosia, 27 October–2 November: 5.

United Nations Secretary-General Reports (1964–93) on the United Nations Operation in Cyprus.

5

EUROPEAN BORDERLANDS

International harmony, landscape change
and new conflict

Julian V. Minghi

INTRODUCTION

The long period of co-operation among neighbours in Western Europe
and of enforced hegemony in Eastern Europe, together with the delicate
Cold War balance between East and West, had until 1989–90 given
post-World War II Europe stable boundaries. The regions created by
their frontiers (the borderlands of Europe) have consequently developed
and, with a few exceptions, have become landscapes of co-operation
and harmony, a process studied by the present author (Minghi 1991).
These developments have been in sharp contrast with the traditional,
more conflict-prone nature of Europe's borderlands. In the late 1930s
Richard Hartshorne (1938) had found no fewer than fifty-six boundary
disputes, and Gordon East (1948) clearly documented the extent and
impact of the many European boundary rearrangements after the war.

The momentous events of the past two years – the withdrawal of the
Soviet Union as a Cold War adversary to the West; the consequent
collapse of its domination of the states of Eastern Europe; the internal
changes in these former satellite socialist states – all necessitate a basic
re-evaluation of the contemporary status of Europe's borderlands,
especially those lying along the former East–West divide, and those
running between Eastern European neighbours. Furthermore, any such
re-evaluation raises serious and worthwhile questions about the likely
changes expected in the human landscape of these borderlands for the
decade of the 1990s.

The aim of this chapter is not to make any such systematic re-
evaluation or to predict the future but rather to focus on contemporary
European borderlands with an emphasis on the process of change in
Western Europe over the past few decades, and on the implications this

89

may have for the future of Europe's borderlands in the light of these revolutionary events. For it seems, without doubt, that the structural breaking-up of the political geographic *status quo* of over forty years' duration is already bringing a level of instability sufficient to generate new conflicts which could, if not actually reform boundary locations, at least reshape the geography of many European borderlands.

First, a case must be made for developing a focus on borderlands as opposed to one on boundaries themselves. Following this, evidence will be presented from several Western European cases concerning the nature of this process of landscape change over time as harmony and co-operation have replaced conflict and hostility, especially where neighbours have found it convenient and politic to use common borderlands as symbols of and investments in their new harmonious relationships. In turn, the relevance of this evidence will be evaluated against some of the contemporary developments both within Eastern European borderlands and in the borderlands between East and West.

THE CASE FOR BORDERLANDS

International boundaries mark the edge of national space and are also interfaces between states. As such, they create their own distinctive regions, making an element of division – the boundary – also a vehicle for regional definition. Hence the concept of 'borderland'. Boundary-dwelling characteristics become dominant moulders of the cultural landscape, falling within the shadow thrown by the boundary. These characteristics disappear as one moves in either direction away from the boundary into the territorial domain of the states divided. Thus the borderland provides a unique opportunity for a local-scale regional focus but within an international context. It can also become a very readable barometer of changes in the relations between the polities involved, especially when studied in a temporal setting.

Most helpful in defining the borderlands concept is the operational model of transaction flows across boundaries presented by John House (1981) (Figure 5.1). House aggregates cross-boundary flows in space and time. Locally generated Az–Bz borderlands transactions are virtually non-existent under the stress of confrontation and the threat of conflict between neighbours, when A–B transactions dominate. On the other hand, as neighbours evolve from such a state of crisis to a more stable and harmonious relationship, A–B transactions improve and a sharp rise occurs in Az–Bz transactions, both in number and import-ance. The core of the case for borderlands studies thus rests on the

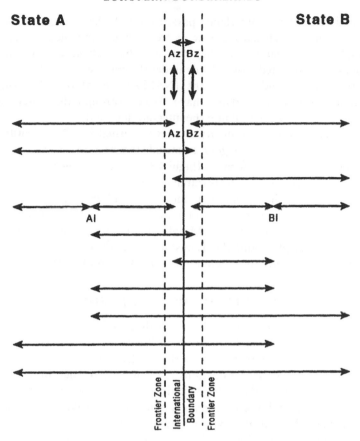

Figure 5.1 Operational model of transaction flow across boundaries
Source: House 1981, p. 295
Note: Az, Bz frontier zones
Ai, Bi regional (provincial) centres
A, B national (federal) centres

assumption that in regions where the pendulum swings rapidly from conflict to co-operation among states – such as the situation along the former 'Iron Curtain' in contemporary Europe – a most fruitful focus for understanding these changes rests in the landscape manifestations provided at the local level interfaces – the borderlands. In the first volume of this series, Oscar Martinez examines in detail this process of borderlands change from one of conflict to co-operation in the case of the boundary between Mexico and the United States. For over three hundred years these borderlands passed through three stages of evolution, at first as 'alienated', next as 'co-existent' and then as 'interdependent' borderlands, and they may now be evolving into a fourth stage as an 'integrated' borderlands under the framework of the North American Free Trade Agreement (NAFTA). In Martinez's terms, Western European borderlands have flashed through these steps in forty-five years and are well into the integration period.

A review article on boundary studies published in 1963 identified borderlands as a special category, to include those studies with an areal, as opposed to a linear, focus (Minghi 1963). More recently, Prescott (1987) identified, as a major research concern for the study of boundaries, a focus on the boundary as an element influencing the cultural landscape of the border region; and in the introduction to their book on border landscapes, Rumley and Minghi (1991: 1–14) trace the evolution of the concept of borderlands in studies over the past fifty years and argue for its relevance to contemporary boundary study (295–8).

EVIDENCE FROM WESTERN EUROPE

Ample evidence can be found in the analysis of the evolution of the cultural landscape of the borderlands of Western Europe over the past four decades to support the notion that under 'normal' peaceful conditions, House's Az–Bz transactions take on a major role. In my own longitudinal study of the Alpes Maritimes region between France and Italy, I found that dominant elements moulding the cultural landscape had changed dramatically over twenty-five years (Minghi 1981). The situation identified by House (1959) – a decade or so after a major post-World War II locational change of the boundary in 1947 favouring France at the expense of its war enemy Italy – had disappeared and the cultural landscape had changed dramatically by the early 1980s. House had found that national policies, without reference to the well-being of the borderlands communities, dominated in the still-hostile setting between Italy and France. Out-migration, especially of the Italian element of the population, was heavy. The new boundary had broken up the 'ecological' balance of many borderland alpine communities, causing bitterness and decline. The aggressive exploitation of the region's hydro-electric power potential by France disrupted the already fragile cultural patterns of the region. Mandatory conscription into the armed forces of France, which attempted unsuccessfully to retain its crumbling colonial empire, led to heavy losses of young males in the French segment of the borderlands population, while ironically Italy, with no colonies left to fight over, required only a short twelve-month NATO military obligation. A railway, completed in 1929 and linking the region with the Po Valley to the north and the Italian Riviera at Vertimiglia to the south, was destroyed during the war and, despite a clause in the 1947 Peace Treaty allowing for its reconstruction, remained in ruins through French intransigence. Certainly, this was still a landscape of conflict in the late 1950s.

Now conditions of harmony and co-operation prevail and the cultural landscape is moulded by a new set of factors. A transborder agency, AlpAzur, has sprung up to assist local governments in solving problems of the borderlands scale and to articulate the particular needs of the borderlands communities to the massive and insensitive bureaucracies of France and Italy. Relations between Italy and France, both EEC founder-members in the Treaty of Rome, are close. Indeed, it has been their desire to use the borderlands as a symbol of this harmony that is, at least in part, the cause of a new set of problems that the author maintains have become typical in Western Europe's borderlands, and may very well eventually develop in the rest of Europe. The railway has been rebuilt and has improved access to the region. An 'international' park – actually, two adjoining national parks combined – has been created from the borderlands' high alpine regions. Yet it is precisely these two major events, so symbolic of co-operation among Western European states, that have paradoxically generated conflict at the local level. The rebuilt railway line is operated at high cost and hence passenger services are minimal, with trains scheduled more to a daily symbolic link between Turin, Nice and Ventimiglia than to convenient commuting of the borderlands' population to jobs outside the region in both countries. Local and regional transportation facilities in border regions sometimes enhance, but frequently inhibit, the integration of labour-market areas in Europe. The preservation of pristine alpine environments for future generations of Europeans conflicts directly with the perceptions of the local community, French and Italian, as to its future economic development and well-being. The locals see their traditional hunting practices and potential for winter-sports development being eliminated.

Leimgruber (1989 and 1991) has shown the increasing importance of understanding the distinctive perceptions of communities in borderland locations in Western Europe, and that these perceptions identify the populations either side of a boundary as ever more similar to and empathetic with each other, and at the same time in growing contrast to populations outside the borderlands proper. It is not surprising, therefore, that issues of significance in borderlands are now less drawn along national lines but more on local and regional interests across the boundary. Two other Western European examples of the evolution of harmonious borderlands are worth consideration. The issue of the Italian South Tyrol, since 1919 a festering sore in relations between Italy and Austria, seems now, after twenty years of genuine cultural autonomy, to have been healed. Indeed, in response to this return to health of Austro-Italian relations, Italy is now taking the lead in sponsoring Austria for membership in the EC, while Austria has under consideration legislation which will renounce once and for all any claim to the South Tyrol. The Swiss industrial city of Basle, for so long restricted by the confrontation between France and Germany, its two suburban neighbours on the north side, is now blossoming under the aegis of perhaps the most successful of Europe's score of transborder agencies, the Regio Basiliensis, under whose influence a tri-national facility has been built on French territory, the Basle–Mulhouse–Freiburg International Airport.

Western European borderlands are no longer regions of national confrontation dominated by security considerations and the potential for war. Conflicts remain, however. Issues now tend to be about local and regional control in which both segments of the borderlands often team together in their own self-interest as a borderlands 'community' to protect themselves against national policies which, in the spirit of international co-operation, can work to their detriment.

TRENDS IN THE NEW EUROPE?

It remains somewhat premature to identify with any precision just how relevant this Western experience of the evolution of close-knit borderlands communities will be for Eastern Europe, but there is room for speculation. The inner-German border disappeared in October 1990 but was so entrenched (Ante 1991) that it is likely to retain cultural-landscape expressions for decades. Indeed, if plans to retain some segments as reminders of this sorry period of German history are carried through, a relic border landscape may also be preserved, although other segments are undergoing great changes (see Buchholz, Chapter 2).

Let me take two contemporary examples as possible indicators for future borderlands problems in Europe. The events in the former Yugoslavia since mid-1991 serve to underline the importance of borderlands and to provide a glimpse of at least one type of borderlands crisis likely to face Europe in the 1990s. The unilateral declaration of independence of Croatia and Slovenia in late June 1991 precipitated some ominous developments for countries threatened by internal break-up based on ethnic nationalism. Though small in area and population, somewhat eccentric to the Zagreb–Belgrade core of Yugoslavia, and with a generally mono-ethnic population, Slovenia borders on Italy, Austria and Hungary. With the territorial integrity of the state of Yugoslavia under threat from Slovenian independence, the Yugoslav federal forces gave highest priority to securing border crossing-points and the two international airports of Ljubljana and Maribor in June and early July 1991. When federal forces garrisoned in Slovenia ran into serious difficulties with Slovenian militia in attaining this goal, the problem of access to ground reinforcements became obvious – the necessity of crossing long distances by road from Serbia through Croatia, and then through Slovenia to these points in the borderlands with the West. While visiting north-western Slovenia in May 1991, the author was assured that this idyllic alpine border region was so distant from the areas of conflict between Serbs and Croats that it would be immune from Yugoslavia's internal problems!

The *force majeure* of central governments may not be able to change people's minds in border areas, but a priority in retaining the territorial integrity of the state creates actions that guarantee immediate involvement of international border regions as well as internal borders in any 'break-up' conflict. In turn, neighbours with civil war literally on and even spilling over their doorstep cannot remain detached. The steady evolution over decades of borderlands from a state of conflict to one of harmony can be abruptly interrupted and set back. For example, the Italo-Yugoslav borderlands, since the settlement of the Trieste Zones A–B question in 1953, has evolved through such a period (Klemencic and Bufon 1991); yet when Slovenia and Croatia declared independence from Yugoslavia these borderlands faced a crisis of territorial control. The implications for borderlands of other countries facing similar problems of nationality groups demanding autonomy are clear. Already embarrassed earlier in 1991 by its inept handling of the Albanian refugee crisis from across the Adriatic, Italy placed its troops in June 1991 on full alert along its border with Yugoslavia and was fearful it could face a wave of refugees, especially from among the roughly 200,000-strong minority Italian population of Istria and Dalmatia if conditions in Croatia deteriorated. Significantly, the Italian Foreign Affairs minister was a leading force in the EC's role as a peacemaker in June 1991.[1]

From the post-war Western experience, the EC seems to favour maintaining the *status quo*, but not by force. This has been highlighted in the EC reaction to the Croat–Serb and Bosnian conflicts, where the Community has maintained a policy of demanding no forced changes to borders, but has been reluctant to use military action to support this position. Negotiations and compromise in satisfying ethnic demands are favoured. Yet as Klemencic and Bufon (1991) point out, the harmonious development along the Italo-Yugoslav borderlands is due less to rational policies of co-operation than to the growth of House's Az–Bz transactions, and these by their very nature link Venezia-Friuli Italians with Slovenes and Croats in a growing borderlands community. Indeed, the growth of anti-Rome regionalism within Italy increases the complexity of this issue. The 'leghe' movement in Northern Italy is seen as a growing threat to the unitary system of governance, as the Lombard, Piedmont and Veneto Leagues – the last-named adjoining Croatia and Slovenia – expand their regional bases of power. By the same token, sympathy for Slovenian independence ran much more deeply in the neighbouring province of Carinthia than at the Austrian national level. Hence the policy conflicts at different scales, already identified as part of the evolution of Western Europe's borderlands, could well be of fundamental significance in the Yugoslav case and in other similar cases in Eastern Europe.

The Austro-Hungarian borderlands provides a telling example of change along the old East–West Cold War boundary. As part of the Iron Curtain, it was a region of minimal contact, with Hungary erecting barriers to its own people moving West, broken temporarily and dramatically only during the 1956 Revolution when tens of thousands of people fled to Austria. In the late 1980s Hungary declared it would no longer respect the Warsaw Pact's restrictions on human movement westward and thus hastened the collapse of other communist regimes in Eastern Europe, as flows through this 'back door' exit reached flood stage. Hungary then scrapped its requirement for exit visas for its own citizens and, by a creative and symbolic act, made the destruction of its Cold War boundary fortifications 'projects of joy' by local borderlands communities – Hungarian and Austrian – in 1989. Ironically, only two years later, there was a sharp increase in efforts by Austria aimed at restricting entry at crossing-points and there was a comprehensive deployment of large segments of the Austrian army to patrol the border zone to intercept and to return people entering illegally. Austria's fear, as a front-line state, of inundation by migrants from Eastern Europe and the Soviet Union has served to generate this seemingly paradoxical situation, one which is becoming typical along the former Cold War borderlands of Europe.

Within communist Europe, internal movement was restricted by each country requiring exit permits from its own citizens. With democratization, this brake to human movement has been removed and yet the number of border crossing-points and associated facilities remain unchanged. In turn, this has led to chronic congestion at crossing-points, generating chaos and international crisis as states act to stem the inflow into and often through their country of unwanted goods and people by placing restrictions on entry from 'fraternal' neighbours for the first time in forty years.

CONCLUSIONS

Soon after World War II, Gordon East (1948) quoted Winston Churchill from *The Times* in January 1948 as asking rhetorically 'who can believe that there will be permanent peace in Europe ... while the frontiers of Asia rest on the Elbe?' As of October 1990 the Elbe became again an internal river within a reunified Germany and hence is a European relic boundary. After entering its fifth decade, the 'Asian' Russian presence is now virtually gone. While the absence of war during this period was hardly permanent peace, by the same token there is no guarantee of 'permanent' peace in the new Europe. Indeed, East's own prediction was not so far off when he wrote 'we may come reluctantly to believe that the Iron Curtain is potentially a safety curtain, and that its preservation and a bisected Europe offer the only hope for peace in our time' (East 1948: 33). As it comes out of the deep freeze of over four decades under *Pax Sovietica*, Eastern Europe is passing through a period of adjustment in which change is necessary and inevitable. Certainly, the Western European experience of peaceful evolution within stable boundaries places a powerful influence on any process for radical change, as the role of the EC in the Yugoslav crisis indicates. However, the Western European model for the evolution of borderlands is less likely to become as broadly relevant within Eastern Europe as it is at the East–West interface borderlands, where the pressures are greater. In any case, the dynamics of the new Europe assure us of a lengthy period of adjustment – certainly for the rest of this century – in which Europe's borderlands seem destined to play a central role and hence deserve attention for intensive and comparative study.

NOTE

1   Since mid-1991 the conflict within former Yugoslavia has developed

dramatically. It took Slovenia and Croatia six months to achieve recognition by the international community, and the war has developed from a conflict over secession to one of territorial control between Croats, Serbs and Muslims. Though the boundaries of the former Yugoslavia in the north (between Slovenia and its neighbours Italy, Austria and Hungary, and between Croatia and Hungary) are not a source of conflict, there are real fears that Serb expansionism could threaten borderlands to the south, particularly in Kosovo. Ethno-territorial strife within Croatia and Bosnia-Herzegovina have generated new *de facto* border problems and the likelihood of 'alienated' borderlands for years to come.

REFERENCES

Ante, U. (1991) 'Some Developing and Current Problems of the Eastern Border Landscape of the Federal Republic of Germany: The Bavarian Example', in D. Rumley and J.V. Minghi (eds), *The Geography of Border Landscapes*, London: Routledge.

East, W.G. (1948) 'The Political Division of Europe', Inaugural Lecture, London: Birkbeck College.

Hartshorne, R.H. (1938) 'A Survey of the Boundary Problems of Europe', in C.C. Colby (ed.), *Geographic Aspects of International Relations*, Chicago: University of Chicago Press.

House, J.W. (1959) 'The Franco-Italian Boundary in the Alpes Maritimes', *Transactions IBG*, 26: 101–31.

—— (1981) 'Frontier Studies: An Applied Approach', in A.D. Burnett and P.J. Taylor (eds), *Political Studies from Spatial Perspectives: Anglo-American Essays on Political Geography*, New York: Wiley.

Klemencic, V. and Bufon, M. (1991) 'Geographic Problems of Frontier Regions: The Case of the Italo-Yugoslav Border Landscape', in D. Rumley and J.V. Minghi (eds), *The Geography of Border Landscapes*, London: Routledge.

Leimgruber, W. (1989) 'The Perception of Boundaries: Barriers or Invitations to Interaction?', *Regio Basiliensis*, 30/2/3: 49–59.

—— (1991) 'Boundary, Values and Identity: The Swiss–Italian Transborder Region', in D. Rumley and J.V. Minghi (eds), *The Geography of Border Landscapes*, London: Routledge.

Minghi, J.V. (1963) 'Boundary Studies in Political Geography', *Annals AAG*, 53: 407–28.

—— (1981) 'The Franco-Italian Borderland: Sovereignty Change and Contemporary Developments in the Alpes Maritimes', *Regio Basiliensis*, 22/2/3: 232–46.

—— (1991) 'From Conflict to Harmony in Border Landscapes', in D. Rumley and J.V. Minghi (eds), *The Geography of Border Landscapes*, London: Routledge.

Prescott, J.R.V. (1987) *Political Frontiers and Boundaries*, London: Allen & Unwin.

Rumley, D. and Minghi, J.V. (eds) (1991) *The Geography of Border Landscapes*, London: Routledge.

Part II
ASIA–PACIFIC

6

THE IMPACT OF RIVER CONTROL ON AN INTERNATIONAL BOUNDARY

The case of the Bangladesh–India border

Nurul Islam Nazem

INTRODUCTION

The aim of this chapter is to highlight how a geographically dis-advantaged and economically weak country co-exists with a neighbour that is larger, economically more powerful and geographically more advantaged. South Asia provides a unique situation to examine such a case. The centrality of India in South Asia provides her with an edge over her weaker neighbours in her foreign-policy dealings and in strategic matters. The comparatively weaker and geographically periph-eral nations of South Asia face external vulnerabilities *vis-à-vis* their stronger neighbour (Hafiz 1989). In the case of Bangladesh, for reasons of its geopolitical situation, most of its perceived threats originate from the policies of India. Part of the problem lies in the fact that India virtually surrounds the country from all sides except to the south.

The relations between Bangladesh and India, which are virtually dictated by this geopolitical condition, have a chequered history. The relations between the two countries speedily developed into a level of cordiality immediately after the independence of Bangladesh in 1971 and turned into mutual suspicion and mistrust in the wake of the change of government in Bangladesh in 1975 (Iftekharuzzaman 1989). There are a number of unresolved geopolitical issues which are the main source of discord between the two countries. First and foremost is the knotty problem of sharing waters of the common border rivers; second, the cross-border immigration problem, particularly in the hill districts of

101

Chittagong; third, delimitation of the maritime boundary; and finally the emerging boundary problems due to the shifting of shared river courses. The focus of the present chapter is on the last issue, that is, the problems of boundary demarcation which have mainly arisen from the Indian policy towards the border rivers between the two countries.

The chapter is divided into three sections. The first section outlines the geopolitical situation of Bangladesh; the second briefly focuses on how this situation is exploited by India and poses a serious problem to the management aspects of the country's river boundary. The final section briefly touches on the implications of the exploitative approach of India toward Bangladesh, her smaller neighbour.

GEOPOLITICAL SETTING OF BANGLADESH

Bangladesh is mainly a flood plain containing a combined delta of the Ganges, the Brahmaputra and the Meghna, is bounded on the west, north and east by India, on the south-east by Myanmar and to the south by the Bay of Bengal. Apart from the 278 kilometres of common border with Myanmar and 730 kilometres of seaboard, the rest of the 3,695-kilometre land boundary lies with India (Figure 6.1). With an area of 143,998 square kilometres and over 110 million in population (in 1990), Bangladesh is one of the largest deltas in the world. Due to the

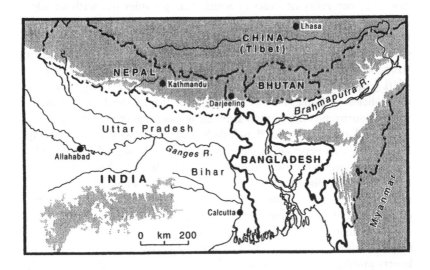

Figure 6.1 Bangladeshi rivers: the basin

102

location of the great Himalayas to the north and the Bay of Bengal to the south, the country is characterized by an effective monsoon climate which produces heavy rainfall. In addition, the surface-water flows and a large and complex network of rivers are a vital component of the geography of Bangladesh. In total, these rivers are at least 24,000 kilometres in length and cover about 9.7 per cent of the country's total area (Bangladesh Bureau of Statistics 1985). Some of these rivers have perennial flows, but most of them are intermittent. The Ganges, the Brahmaputra, the Meghna, the Teesta and the Karnafuli are perennial flows. The catchment areas of perennial rivers, particularly those of the Ganges and the Brahmaputra, are spread over several countries: for example, the Ganges over China, Nepal, India and Bangladesh; the Brahmaputra over China, Bhutan, India and Bangladesh; and the Meghna over India and Bangladesh. In fact, the inflow of waters from India constitutes about 92 per cent of the total surface flow, while only about 8 per cent is generated within the country.

The drainage basins of the largest three rivers in the country, that is, the Ganges, the Brahmaputra and the Meghna, with their innumerable tributaries and distributaries, cover about 80 per cent of the total land area. All these rivers, including other inland channels and streams, present an intricate web of river networks across the whole country (Figure 6.2). Most of these rivers enter directly into Bangladesh territory across the international boundary and some of them form the borderline between Bangladesh and India before entering into its territory. There are some other rivers which, originating in India, enter into Bangladesh and after flowing some distance re-enter Indian territory. There are still some other watercourses which have double entry into Bangladesh. Most of these rivers undergo local physiographic changes either by erosion or by deposition every year. It is not surprising that the border rivers very often cause problems in the management and demarcation of the international boundary between the two countries.

Bangladesh is cut through by these countless watercourses and the whole range of her economic, social and cultural life has emanated from and been sustained by the historic uninterrupted flows of these rivers since time immemorial. The flat topography of Bangladesh, although ideal for agriculture, is prone to frequent flooding in the monsoon period and shortage of water in the dry season under the prevailing rainfall pattern and river regimes. Although modern technology has opened up an unprecedented opportunity to harness these resources, the potential is constrained by the political process. The boundary between the countries that evolved through political process does not follow the

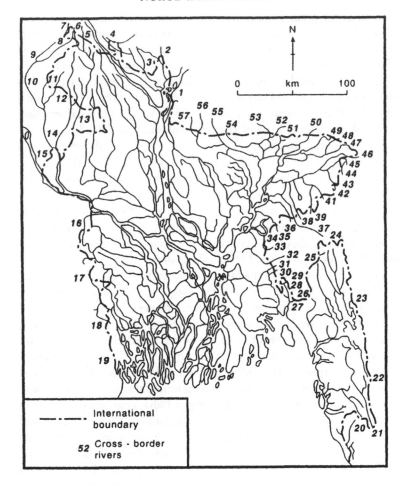

Figure 6.2 Indo-Bangladeshi cross-border rivers
Source: Nazem and Kabir 1986, (and see *Annex*, p. 110, for names).

logic of physical geography. There are at least fifty-four rivers in Bangladesh (JRC 1985), including the three major ones mentioned above, whose head waters are located outside her territory; Bangladesh, being lower-riparian, has hardly any control over them. This presents classic hydro-political problems for the state.

THE CAUSES AND CONSEQUENCES OF BOUNDARY CHANGES

The geopolitical situation outlined above gives India an edge over Bangladesh. The latter country, being lower-riparian, has remained in a disadvantaged position with regard to the management of her water-courses. India manipulates this situation in her geo-strategic calculations and attitude towards her neighbour. This is reflected in India's policy of delaying the resolution of problems on various grounds – for instance, delays in the execution of agreements or in responding to correspondence (or not responding at all) – and, at the same time, gearing up her own strategy without consideration of the impact on Bangladesh (Nazem and Kabir 1986). Bangladesh, on the other hand, continues to be affected by India's river diplomacy which has some far-reaching adverse consequences on her territory. For instance, India's structural measures on various border rivers are mainly designed to withdraw waters, but some of them pose serious international boundary problems (JRC 1981a, 1981b). These activities are not limited to the big rivers, but also affect many smaller common rivers.

Among all the issues between Bangladesh and India, the water-sharing issue is the most pressing and devastating. In 1974 India constructed a barrage on the Ganges river, just 18 kilometres upstream from the Bangladesh boundary in a place called Farakka. Ever since the Farakka barrage was constructed the flow of water in the Ganges has fallen drastically, much to the detriment of the interests of Bangladesh (Nazem and Kabir 1986). Although India made a commitment that it would ensure a mutually acceptable solution to the water-sharing issue before the barrage was commissioned, the solution seems to be as elusive as ever. The negotiations are still going on, and in the process of negotiations Bangladesh is gradually being squeezed by mounting social, economic, ecological-environmental and political pressures as a result of continued diversion of waters not only from the Ganges but also from other border rivers. On the other hand, Bangladesh has been desperately struggling for a long-term solution to the problem during the last two decades.

The process of Bangladesh's struggle and India's reaction is beyond the scope of this chapter. The point that is relevant here is that the river Ganges has been experiencing severe downstream morphological changes since the commissioning of the Farakka barrage which have caused frequent changes in the international boundary between Bangladesh and India (*Holiday*, 4 July 1986; *Bangladesh Times*, 30

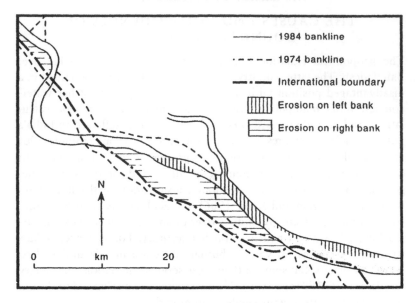

Figure 6.3 The shifting course of the Ganges
Source: Bangladesh Times 1985.

December 1985). Fifteen years ago the Ganges had only one entry-point into Bangladesh, but now it enters at two different points, having a curved course of about 12 kilometres inside Indian territory. This is mainly due to diversion of silt-free water upstream pushing a large quantity of silt into the river in Bangladesh (Nishat 1986). Shifting of river courses due to erosion and sedimentation creates manifold problems not only in boundary management, but also in irrigation, navigation, human settlements and so on (Figure 6.3).

Similar problems are found in the case of the rivers Brahmaputra and Teesta. The river Brahmaputra is one of the largest rivers in the world, with a length of 2,900 kilometres, flowing through several countries. Bangladesh is heavily dependent on its water for various important uses like irrigation, navigation, checking salinity intrusion, fisheries and forestry, and so on. The degree and extent of dependence on this river is much greater than the others because of its large volume of water; but the Indian water-development activities and schemes on the river Brahmaputra in the upstream sections are increasingly posing threats to the economy and ecology of the lower-riparian Bangladesh. Such schemes involve 3,830 kilometres of embankments along the main rivers' tributaries, 770 kilometres of drainage channels and forty-four town-protection projects (Khan and Miah 1983).

The Teesta is another mighty river which flows into the northern dry zone of the country, where Bangladesh undertook a large-scale irrigation project in 1958–9; but due to India's construction of the Gajaldoba barrage in the upstream section of the Teesta the Bangladesh project is under severe threat.

Apart from the Ganges, the Brahmaputra and the Teesta, there are many other rivers which are either border or common rivers between Bangladesh and India. These watercourses are also crucial for Bangladesh; but India has constructed various structures on several of the common cross-border small rivers. The objectives of these structures are not only the withdrawal of water but also other regulating measures. India, for example, has constructed dams on the rivers Buri Teesta, Kherua, Sangli and Ghoramara. She has constructed spurs and weirs on the Monu and Chiri and barrages on the Khoai and Gumti. Similar structural measures were undertaken on many other rivers (JRC 1981a). Due to these activities upstream Bangladesh incurs a considerable loss in terms of life and property. Bangladesh faces erosion problems with wet-season floods in the rivers Kushiyara, Muhuri, Feni, Khowai and Ichamati.

The mid-stream of the river Kushiyara at Zakigonj demarcates the border between Bangladesh and India. On the left bank in India, just opposite to Zakigonj town, is Karimgonj town. The Kushiyara river flows in a meandering course and Zakigonj is situated on the concave bank. The erosion problem of the river Kushiyara at Zakigonj has been seriously aggravated due to the construction of a number of protective measures by India on the left bank of the river since 1973–4 (JRC 1981b). As a result, the deep-water line has been shifting to the Bangladesh side, prompting a debate on the demarcation of the inter-national boundary between the two countries. The problem was discussed at the fourteenth and fifteenth meetings of the Joint River Commission (JRC) in 1985, but the government of India went ahead with the construction of many such protection works, defying the decision of the JRC that pending an agreement none of the parties should undertake any project.

The watercourse of the river Muhuri has been shifted towards the right bank in Bangladesh due to construction, extension and advance-ment of spurs and bank-protection works along the left bank of the river in the Indian territory. A big *charland* of about 200 acres was formed well within the Bangladesh territory but this land is cultivated by

Indians. When the matter was taken up by the concerned authority to their Indian counterpart, the Indian side did not even bother to reply to correspondence (JRC 1985).

The mid-stream of the river Feni at Ramgarh forms the international boundary. The river Feni flows more or less straight between Ramgarh town in Bangladesh and Sabrum town in India; but due to construction of spurs, there has been an erosion problem on the Bangladesh side. The problem was discussed at the JRC meetings but India encroached upon the river by constructing further works and diverted the flow to the Bangladesh side of the Feni river, which is against the joint decision of the appropriate authorities of both countries (JRC 1981b). The same is applicable in the case of the Ichamati, Gomti and Khowai rivers. The Bangladesh government had undertaken a project on the Khowai river with the hope of getting financial support from the Asian Development Bank (ADB); but when the ADB made a reference to the government of India on the project, India intimated her plan for utilization of the water upstream. This resulted in the deferring of the project to be financed by the ADB until an understanding is reached between the governments of Bangladesh and India on the availability of waters for the project (Nazem and Kabir 1986).

CONCLUSIONS

In the light of the above-mentioned discussions two observations on the problems of cross-border river management between Bangladesh and India seem critical. First, various structural activities on the Indian side of the border rivers cause devastating effects on the Bangladesh side. The worst effect is bank erosion, which not only creates problems of boundary demarcation between the two countries, but also displaces thousands of people from their land and shelters. Subsequently, when charlands emerge as a result of siltation, it leads to disputes on the question of ownership and sometimes to armed clashes between the security forces of the two countries. Second, Bangladesh is being deprived of its legitimate share of waters due to upstream withdrawal from even the smaller rivers, which are vital for agricultural activities during the lean period in a food-deficient country like Bangladesh. Besides the water-sharing issue from the large rivers, on which negotiations are underway, the problems with the small border rivers are becoming increasingly acute. The JRC has already been taking up the cases of six such small-river problems besides the major-river issues.

Bangladesh is constantly trying to make India recognize the adverse effects her water politics and her unilateral structural activities on the common rivers are having on Bangladesh; but India has never seemed sincere in its attempts to solve these crucial problems. This is evident from the dialogue between Bangladesh and India during the last two decades at various levels, which has in effect ended without producing tangible results (Abbas 1985). India seems to have followed a policy of presenting a *fait accompli* in dealing with Bangladesh.

Rivers are commonplace in Bangladesh and are a significant feature of the physical and cultural landscape of the country. Proper management of these rivers is of vital importance to national development. Yet management possibilities are severely constrained by external factors, which originate from the divergent Indian position not only on the water sector, but also on the overall geopolitical situation of the country. There can be little doubt that India exploits Bangladesh's geopolitically vulnerable situation to pursue her regional politico-strategic goals. The magnitude of the problem with regard to international-boundary management, however, may not be to the forefront yet, because of the priority that the water-sharing issue continues to hold, it is to be expected that the cumulative impact of all these activities on Bangladesh will give her severe problems in the not-too-distant future.

REFERENCES

Abbas, B.M. (1985) 'Agreement on the Ganges', paper presented at the 'Regional Symposium on Water Resources Policy in Agro-Socio-Economic Development', Dhaka.

Bangladesh Bureau of Statistics (1985) *Statistical Yearbook of Bangladesh*, Dhaka.

—— (1990) *Statistical Pocket Book of Bangladesh*, Dhaka.

Bangladesh Times (1985) Dhaka, 30 December.

Hafiz, M.A. (1989) 'South Asia's Security: Extra Regional Inputs', *BIISS Journal*, 10/2.

Iftekharuzzaman (1989) 'Political Instability, External Vulnerability, Under-Development: The Vicious Circle for Bangladesh', paper presented at the international seminar on 'Development Dynamics: Political and Security Dimensions', Dhaka.

JRC (Indo-Bangladesh Joint Rivers Commission) (1981a), *Working Paper on Various Border/Common Rivers Problems between India and Bangladesh*, vol. I, Dhaka.

—— (1981b) *Working Paper on Various Border/Common Rivers Problems between India and Bangladesh*, vol. II, Dhaka.

—— (1985) *Record of Discussions of Indo-Bangladesh Joint Rivers Commission*, Dhaka.

Khan, A.H. and Miah, S. (1983) 'The Brahmaputra River Basin Development', in Munir Zaman *et al.* (eds), *River Basin Development*, Dublin.

Khatri, K.S. (ed.) (1987) *Security in South Asia*, Kathmandu: Centre for Nepal and Asian Studies, Tribhuvan University.

Mirza, M.Q. (1986) 'The Scourge of Farakka', *Holiday*, Dhaka, 4 July.

Nazem, N.I. and Kabir, M.H. (1986) *Indo-Bangladesh Common Rivers and Water Diplomacy*, BIIS Papers 5, Dhaka.

Nishat, A. (1986) 'Delay in Water Resources Cooperation: A Bangladeshi Perspective on Human Opportunity Costs', paper presented at the conference on 'The Opportunity Costs in Delay in Water Resources Cooperation between Nepal, India and Bangladesh in the Ganges and the Brahmaputra Basins', Gold Coast City, Australia (Institute of Management Sciences Conference, XXVVII).

ANNEX

(Names of rivers numbered in Figure 6.2)

1	Brahmaputra	30	Little Feni-Dakatia
2	Dudhkumur	31	Gumti
3	Dharla	32	Salda
4	Teesta	33	Bijni
5	Kharkharia-Jamuneshwari	34	Howra
6	Talma	35	Anderson Khal
7	Karotoya-Atrai	36	Sonai
8	Dahuk	37	Sutang
9	Mabananda	38	Khowai
10	Nagar	39	Karangi
11	Kulik	40	Lungla (Gopia Lungla)
12	Tangon	41	Dhalai
13	Punarbhaba	42	Monu
14	Pagla	43	Juri
15	Ganges	44	Sonai-Bardal
16	Mathabhanga	45	Kushiyara
17	Kobodak	46	Surma-Meghna
18	Ichamati-Kalindi	47	Sari-Gowain
19	Raimongal	48	Piyan
20	Matamuhuri	49	Dhala (Dhalai Gong)
21	Rankhiang Khal	50	Umium (Bogra)
22	Thega or Kawrpin	51	Nowagang
23	Karnafuli	52	Dhamalia
24	Kasalang	53	Jodukata
25	Myani Khal	54	Someshwari
26	Halda	55	Nitai
27	Feni	56	Bhogai-Kangsha
28	Muhuri	57	Chilla Khali
29	Selonia		

7

PEACE AND CONFLICT IN THE THAI–MALAYSIAN BORDER REGION

Dennis Rumley

INTRODUCTION

This chapter analyses the causes of conflict in border regions of contiguous states using the Thai–Malaysian example as a case-study. It is suggested that one of the principal causes of conflict in border regions is the failure to maximize the primary functions of the political organization of space – participation, representation and resource allocation – on the part of border populations. Five main approaches to a consideration of border-region conflict – positivist, political development, colonial boundary, ethno-religious nationalism and centre–periphery – are discussed, and it is argued that no one approach possesses theoretical or practical pre-eminence. Rather, the chapter argues that expressions of conflict in border regions, especially in the developing world, can be best understood via a synthesis of these various approaches within an overall centre–periphery framework.[1]

PEACE AND THE POLITICAL ORGANIZATION OF INTERNATIONAL SPACE

One of the principal aims of this chapter is to initiate the development of a geographical framework for the deeper understanding of conflict in international border regions. This view is at variance with the trend which has been recently noted in international boundary inquiries away from a concern with conflict and towards a consideration of co-operation (Minghi 1991). It is suggested here that such a trend represents above all an ethnocentric 'Western' orientation which applies primarily to new and emerging boundary arrangements designed principally for economic purposes in the differing contexts of the developed

111

'North'. In addition, what the trend tends to underplay or ignore are the fundamental social, economic and political changes which are in progress in the developing 'South' which are the product of a complex set of interrelated factors (Nyerere 1990).

First, such changes are associated in part with the interrelationship between development, social justice, environmental stress and conflict (Brundtland 1990: 334–51). In this regard, one of the measures by which the success of development might be determined is the degree to which it is able to bring about social justice and thus eradicate negative conflict and bring about peace (Rumley 1991a).

Second, the changes are also clearly linked to questions of political development related to emergent nationalism often suppressed or controlled by colonial or neo-colonial policies and practices carried out either by former European or by current intra-state ethnic elites. Such 'emergent nationalisms' invariably are located on the periphery of developing states along or astride international political boundaries.

Third, it has been noted that there has been a tendency on the part of writers on international relations to ignore race as an ingredient in world conflicts (Tinker 1990: 43). However, in the latter part of the twentieth century it is especially important to emphasize the fact that, despite all the hopes of Marxist colleagues and others, ethnic and religious questions have refused to disappear. On the one hand, we have seen the perpetuation of a 'radical expectancy' in the literature which has assumed that differences in religion and ethnicity would be 'overcome' by class consciousness, even when they straddled international political boundaries. In addition and alternatively, we have seen expressions of a 'liberal expectancy' which assumed that somehow the process of 'modernization' would contribute to homogenization and blur ethnic differences (Wijeyewardene 1990: vi).

All these factors taken together possess fundamental implications for the political organization of space at various scales. From a 'reformist' perspective, the political organization of space should be designed to maximize its essential functions – participation, representation and resource allocation. These functions arise out of basic human needs associated with identity (of self), community (relations with others) and power (control over self). What is especially problematic from the viewpoint of conflict is that this essentially pluralist conception of the state (Johnston 1989) has been in the process of being eroded by an instrumentalist mode of operation. Thus the state functions as much for special-interest groups (for example, ethnic, political-party or class elites, bureaucrats and business people) as much as it does for the resident population, who as a result become increasingly disillusioned and alienated from the political process (Yiftachel, Carmon and Rumley 1991). The 'control' function of the political organization of space becomes increasingly important as a result and is in permanent conflict with the other functions, the maximization of which in any event is highly problematic (Rumley 1991b).

The combination of the emergence of ethno-religious nationalism on the one hand in conflict with the pre-eminence of the control functions on the other has very serious implications for the stability and structure of developing states in cases where conflict is confined primarily to peripheral border regions. This touches at the heart of a fundamental global issue addressed some years ago which can be posed as a theoretical question – what would be the impact of the total removal of all existing international political boundaries and of the process of redrawing boundaries anew using all distinctive large regionalisms as a basic unit? As was noted, if this exercise were carried out, we would have many more political units, the boundaries of which, for the most part, would be significantly different from the present world political map (Knight 1983).

CAUSES OF BORDER-REGION CONFLICT

Analyses of border-region conflicts by political geographers and other social scientists have approached the problem either from a descriptive or from an explanatory perspective. The descriptive perspective, popular among political geographers, has intended to concentrate mainly on the classification of international boundary disputes – territorial, positional, functional and resource-development (Prescott 1990: 98). The explanatory perspective, on the other hand, popular among most other social scientists, has attempted to probe the causes of border-region conflict within a broader theoretical framework. Conflicts either have been examined on a state–state basis – that is, the prime causes of conflict can be attributed to the characteristics of the contiguous states – or can be considered in relation to state structure and to the internal policies of contiguous states. The outcome of the latter is a set of associated conflicts primarily confined to the border region itself.

Overall, the explanatory approach to border-region conflict is under-pinned by a variety of philosophical presuppositions and has been approached from a number of methodological standpoints. In Prescott's view, attempts to date to develop any 'reliable theory' about inter-national political boundaries have largely failed (Prescott 1990: 8). For the sake of analytical convenience, the various approaches to under-standing the causes of border region conflict have been grouped under five main headings – the positivist approach, political development, colonial-boundary impact, ethno-religious nationalism and the centre–periphery viewpoint. In actuality, of course, some of these approaches overlap and converge to a certain extent. Nevertheless, it is argued here that any comprehensive understanding of conflict in border regions in the developing world requires a synthesis of the characteristics of each approach.

The positivist approach

One of the most notable examples of the positivist approach to an identification of the causes of border-region conflict is a study based on data for sixty-six border disputes during the period 1945–74 (Mandel 1980). Mandel's view is that previous social-science research, which has aimed to explain the likelihood of interstate war in relation to geo-graphical distance and with the number of bordering states, is based on questionable assumptions, often yields contradictory conclusions and overall has contributed minimally to an understanding of the causes of border-region conflict (Mandel 1980: 430–1). Five main hypotheses are tested and confirmed during the analysis: these are that the probability of border dispute is highest where (1) contiguous states possess roughly equal levels of power; (2) there are relatively low levels of technology; (3) ethnicity is a factor; (4) the states are members of opposing blocs; and (5) there are two states involved. Clearly, one of the primary difficulties faced by such an approach is that of operationalization in which quantitative indicators are attached to a range of independent and dependent variables for contexts which are invariably highly complex. Thus, the reduction of border-conflict situations to a number of testable propositions will underplay or ignore factors which are difficult to measure. In particular, the 'contents' of space are omitted from any serious consideration. As a result, analyses such as this may have little utility for policy-makers and hence for the resolution of the conflicts being examined.

The efficacy of the positivist approach to peace studies in general has recently been the subject of some debate among political geographers (Taylor 1991: 81). Whereas by some writers a 'mainstream' analytical-empirical viewpoint is preferred (for example, Van Der Wusten and O'Loughlin 1986), to others such a perspective represents an innate conservative orientation which does little to 'legitimize the interests of people' (O'Tuathail 1987). However, the latter radical view is equally problematic from a policy perspective. Furthermore, neither perspective can of itself provide a comprehensive understanding of border-region conflicts in the developing world since the very conceptual basis of boundaries and boundary evolution is fundamentally different from that in the Anglo-American context.

Political development

It has been suggested that the likelihood of conflict within and among states decreases with an increase in the objectively defined level of a state's political development (Lane and Ersson 1989). In particular, the potential for border-region conflict is greatest in contexts of relatively high political instability in an environment of relatively recent nation-building.

The concept of political development is both complex and multi-dimensional, especially in terms of the relationships among its various characteristics. However, knowledge of the multi-dimensional characteristics of the political development of contiguous states can lend some insight into the potential for border-region conflict.

In terms of the present case-study, from the viewpoint of 'political development as democracy' Thailand is currently seen as a 'semi-democracy' (Chai-anan 1989) and Malaysia is perceived to be a 'quasi-democracy' (Zakaria 1989). Objective data across all political-development dimensions for both states (Table 7.1) reveal a different democratic tradition ('democracy'), a relatively low level of social-services expenditure in Thailand ('polity'), a higher military capacity in the latter ('military'), relatively recent nation-building in Malaysia ('institutionalization'), a relatively low level of 'radical'

Table 7.1 Dimensions of political development (after Lane and Ersson 1989)

Dimension	Malaysia	Thailand	Asia	OECD
Democracy	52.2	37.7	43.7	60.1
Polity	55.1	39.8	47.5	57.4
Military	55.1	55.5	61.8	49.1
Institutionalization	63.6	53.4	57.7	40.9
Radical	42.5	41.0	46.1	53.0
Protest	45.4	53.6	46.3	54.6
Violence	57.4	57.0	51.9	48.8

orientation in both states, higher 'protest' in Thailand and an above-average measure of political instability in both states ('violence').

Colonial-boundary impact

Before the process of European colonization, many parts of the developing world possessed different concepts of boundaries and of political units in which territorial control was defined in terms of spheres of influence. Even now, in parts of South East Asia some borderland areas are still in effect frontier zones in which boundaries are often poorly demarcated (Lee 1982). Boundaries were seen as 'convenient markers' rather than as actual 'lines of division'. Thus, 'unauthorized movement' and 'illicit trade' across boundaries must be interpreted in the light of pre-colonial behaviour patterns (Lim 1984: 67). The intervention of European colonial powers created 'discontinuity' on indigenous kingdoms via a process of competition for territorial control and the imposition of boundaries by colonial cartographers without due regard for indigenous society, culture and political structures (Drummond and Manson 1991: 217).

It has been argued that in the case of South East Asia, the primary cause of conflict in border regions was the 'imperfect' definition and delimitation of political boundaries by European colonial powers. It is thus not unusual to find boundaries separating groups such as the Malays in the Thai–Malaysian border region, the Lao in the Thai–Laos region, the Sulu speakers between Sabah and the Philippines and the Papuans separated by the Irian Jaya–Papua New Guinea boundary. All of these regions are areas of potential or actual conflict (Lee 1982: 28–9).

Ethno-religious nationalism

From a Western liberal-democratic perspective, concern has been expressed over the potential for change in the political organization of space and the relations among states of both the 'ethnic revival' and of religious fundamentalism, especially where these two processes coincide. 'Ethnonationalism', for example, has been seen as one of the most significant 'destabilizers in the current world system'. Furthermore, from this viewpoint it is arguable as to the degree to which the so-called 'fire-ring of Islam' has implications for world peace (Kliot 1991: 12–13). The combination of imposed colonial political boundaries, a relatively low level of political development and a significant degree of ethno-religious nationalism clearly reinforces the likelihood of border-

116

region conflict. However, in part this can and will be mediated by the socio-geographical structure as well as policies of contiguous states.

The centre–periphery viewpoint

As has been argued elsewhere, an adaptation of Rokkan's conflict model (Figure 7.1) yields at least four basic dimensions of peripherality – cultural, economic, political and geographical (Rumley and Minghi 1991b: 4–6). From a research perspective, the cultural dimension is essentially concerned with conflicts between elite and minority groups. The economic dimension is concerned with conflicts which arise out of economic exploitation and an uneven distribution of wealth. The political dimension is concerned with conflicts which arise out of variations in political participation and political power, and the geographical dimension is related to distance and perception of strategic territorial advantage as well as with the causes of localized regional conflicts. The latter may well arise as a result of the geographical co-incidence of any one or more of the other dimensions but conceptually is the least developed (Wellhofer 1989).

Clearly, the potential for border conflict is greatest where the four

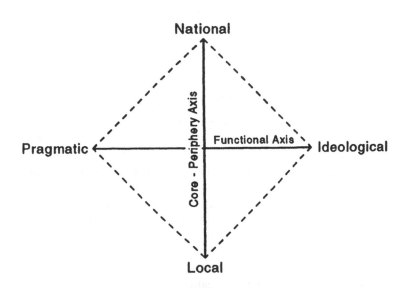

Figure 7.1 Rokkan's conflict model
Source: Taylor and Johnston 1979: 129.

117

dimensions coincide, and such situations demand co-operative government action on behalf of contiguous states. It is likely, especially in the developing world, that inhabitants of border regions tend to enjoy the least political power of any group in the state and participate less. In addition, within the state they are likely to be regarded as culturally and/or economically peripheral. Incomes per head, for example, are likely to be lower and state allocations of economic resources per head are likely to be less. 'Peripheral inhabitants' tend to be more culturally independent and more conservative than those in central locations and are therefore less willing to adapt to a national set of norms and a national culture which, on the one hand, might deprive them and yet, on the other, demand their unqualified allegiance. In such contexts, strong national pressure to fully adopt national norms might well encourage peripheral inhabitants into radical political action (Rumley 1991c).

EXPRESSIONS OF BORDER-REGION CONFLICT

Four categories of border-region conflict are discussed here – positional disputes and border violations, social-cultural peripherality, political peripherality and economic peripherality. Whereas the first is descriptive of state–state conflict, the other three arise primarily from intra-state structures and are primarily explanatory. Before this discussion, however, a brief overview will be provided of the study context – the Thai–Malaysian border region.

The Thai–Malaysian border region

As has been elaborated elsewhere, the case of the Thai–Malaysian border region (Figure 7.2) was chosen for analysis for four principal reasons (Rumley 1991c). First, it was chosen because of the different ethnic and religious structure of each state, their geographical variation and the respective state policies regarding these structures which tend to be highlighted in the border region. For example, the Thai–Malaysian case is the only complete international land-boundary meeting-place of Islam and Buddhism, forming part of an important 'primary religious divide' in the former South-East Asian 'shatterbelt' (Spencer and Thomas 1971: 347).

In particular, there exists a sizeable Malay Muslim minority in the four Southern Thai provinces of Narathiwat, Pattani, Satun and Yala (Haemindra 1976). The Malay Muslim population of Southern

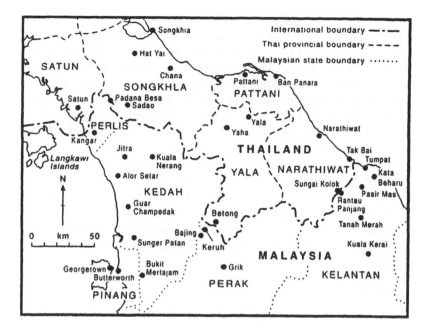

Figure 7.2 The Thai–Malaysian border region
Source: Rumley 1991c: 130.

Thailand was estimated to be in the order of 750,000 in 1962, or about one-fifth of the Malay population in Malaya (Lamb 1968: 170). By 1990 it is estimated that the Malay population in Southern Thailand had risen to 1.5 million compared with an estimated 8.5 million in peninsular Malaysia.

In Malaysia as a whole, Malays comprise about one-half of the total population, although there are significant minorities of Chinese (about one-third) and Indians (about 10 per cent), but there is considerable cultural diversity within these groups (Carstens 1986: 1). There is also a significant geographical variation in the distribution of the groups with four of five of the border states (Terengganu, Kelantan, Perlis and Kedah) having the highest peninsular proportions of Malays, for example. The other border state, Perak, contains significant concentrations of Chinese and Indians, and consequently the Malays are in a minority in that state (Shafruddin 1987: 120). Both Thailand and Malaysia possess an overt nationalist policy based on a pre-eminent ethnic and religious code of behaviour, and in combination these policies have important future implications for the Thai–Malaysian

119

border region. Thailand's policy is based on the pre-eminence of Thai culture and of the Buddhist religion, while Malaysia's policy is based on the pre-eminence of Malay culture and of Islam. These policies interact in the Thai–Malaysian border region with economic-development policies, communal politics and incipient Islamic fundamentalism. Second, the contiguous states possess different state structures. For example, Thailand is a relatively large, highly centralized state with a population of 55.7 million in 1990. There was some evidence to suggest, however, that despite the February 1991 military coup (the seventeenth in modern history), Thailand was soon taking some initial steps to give provincial and local administrations more authority (*FEER*, 6 June 1991). Malaysia, on the other hand, had a population of 17.9 million in 1990 and possesses an elected federal government. In the most recent (October 1990) federal election, questions of race and regionalism were among the most significant issues (*FEER*, 1 Nov 1990).

A third reason for an examination of the Thai–Malaysian case is historical. Unlike Malaysia, Thailand did not experience European colonial rule. In Malaysia British colonial ideas of centralization and uniformity of control were built into a federation which was designed to accommodate the legacy of Malay states and the institution of the Sultanate (the source of some political friction in early 1993) rather than to accommodate communalism or size as in many other federal systems (Shafruddin 1987). In Thailand, on the other hand, the lack of any political alternative to authoritarian rule has been seen as virtually a by-product of the avoidance of colonial control (Cady 1974: 442). The differential legacy of British colonial rule is thus an important ingredient in the understanding of the nature of conflict in the Thai–Malaysian border region.

A fourth and related reason for being concerned with the Thai–Malaysian case derives from the historically strategic importance of the Malay peninsula and its relationship to emergent Thai and Malay nationalism in the twentieth century. This relationship is brought into sharp focus in the Thai–Malaysian border region. It has been noted, for example, that, given its location approximately half-way between the civilizations of India and China, the peninsula has always been a barrier to trade. Partly as a consequence of this factor it has functioned as a region of contact between different peoples for many centuries (Wheatley 1961; Hall 1981).

The Thai–Malaysian international boundary is geographically more peripheral to Thailand than it is to Malaysia. From the federal capital of

Kuala Lumpur, for example, the straight-line distance of the boundary is approximately 300 kilometres. In contrast, the international boundary is about twice that distance from Bangkok (Rumley 1991c).

Positional disputes and border violations

The Thai-Malaysian international boundary was finalized through the 1909 agreement between Siam and Britain and the coincidence of a principally physical boundary with a political boundary was assumed to facilitate the process of delimitation and control. However, there have been a number of ongoing disputes in the central jungle regions and in the western section over the precise location of boundary markers. Overall, on both sides of the boundary security is very tight, but perhaps more so on the Malaysian side. The erection of a large wall by the Malaysians at the Kroh crossing and the walled compound patrolled by Malaysian guards at Pedang Besar are evidence of this. However, Thai border security, in general, appears to increase from west to east and is especially tight in Narathiwat, partly associated with smuggling which is an endemic component of the Thai-Malaysian border economy.

Among the most problematic sections of the boundary in terms of physical delimitation and control is the eastern section between Narathiwat and Kelantan which runs along the Sungei Kolok. There are at least two aspects to this physical problem. First, the determination of the boundary is problematic close to the river mouth since the location of small islands can change both seasonally and yearly. Islands which are visible in the dry season, for example, can disappear in the wet. Furthermore, the location of river-mouth islands from 1955–71, for example, has changed and this makes the identification of the boundary very difficult. These issues have become associated with local conflicts over land use, fishing rights, river pollution and the movement of goods and people (Rachagan and Dorall 1976: 54).

This leads to the second aspect of this problem, which is more evident upstream. In the vicinity of the twin towns of Sungai Kolok and Rantau Panjang, for example, the Kolok river at times is very narrow and easy to cross. Control of river crossings and hence of smuggling is very difficult if not impossible to enforce. Although there is only one 'official' crossing-point, fifteen illegal crossing-points are known to the local inhabitants in Sungai Kolok. Furthermore, in Malaysia and Thailand it is illegal to maintain dual citizenship. However, due to the comparative ease of crossing the boundary and the uncertainty of home location, it is possible for border residents to maintain dual citizenship.

Pattani Malays from Thailand have even been known to vote in Kelantan state elections (Rachagan and Dorall 1976: 55)!

In a region like South East Asia in which the rate of forest clearance and degradation has occurred on a dramatic scale, the question of illegal logging is highly politically sensitive, particularly if it comprises cross-border illegal logging (Hurst 1990: 212). In May 1991, for example, eleven 'officials' from Kelantan State were arrested for illegally entering Thai territory and engaging in illegal logging in a southern Thai border forest in Narathiwat: one of those arrested happened to be the Director of the Kelantan State Forestry Department! It was claimed by Thai provincial authorities that the illegal operation had been going on for some time (*The Nation*, 5 June 1991). On the other hand, the Malaysians claimed that the officials strayed unintentionally across the border without any illegal intent. Some Malaysians also believed that Thailand was using the detainees in order to bargain for concessions for its fishermen in the ongoing fishing dispute.

Socio-cultural peripherality

Of the more than two million Thai Muslims, more than three-quarters live in the south and there has been an ongoing series of disputes between the southern Thai Muslims and Bangkok-based bureaucrats and security officials associated with socio-cultural peripheralization. In general, it appears that Thai state policy regarding the minority to a certain extent has exacerbated regional alienation (Keyes 1987: 131). The Malay-speaking Muslims of southern Thailand do not possess any innate linkage to the Thai nation. This is expressed in terms of a low level of proficiency in the Thai language and in terms of cross-border interaction patterns. For example, residents on both sides of the international boundary have relatives on the other side. Every day, hundreds of children cross illegally from Sungai Kolok in Thailand to Rantau Panjang in Malaysia in order to attend school.

The degree of alienation has been politically expressed in different ways and political conflict between the Thai Muslims and the central government bureaucracy has tended to exacerbate socio-cultural peripherality. On the one hand, the view of many Thai security officials is that any local opposition is yet another manifestation of Shia-sponsored separatism. On the other hand, the view of Thai Muslims is that the local imposition of central policy by the Bangkok-appointed bureaucracy is another example of 'ingrained Thai-Buddhist cultural chauvinism'.

Political peripherality

The socio-cultural peripherality of Thai Muslims is reinforced by the political-geographical peripherality of the Thai–Malaysian border region. Even though there have been some recent moves to decentralize, the problem remains of the extreme centralization of the Thai state. An important perception on the part of local inhabitants is one of 'external control' by an indifferent government. The apparent lack of trust in political relations is due in part to an insignificant local role in local administration. All southern governors and senior staff are Buddhist, as are many local school teachers, some of whom are sent unwillingly to the south from other provinces. There has consequently developed a kind of 'vicious circle' of political peripheralization feeding local desires for greater regional autonomy and even separatist tendencies, especially in Pattani. As part of this 'syndrome' many local inhabitants have been turning to spiritual and cultural values, which in turn has served to reinforce socio-cultural peripheralization. Such a situation contains the necessary ingredients for external exploitation and the sparking of communal unrest. For example, the Thai government is concerned over the increasing numbers of border-region inhabitants who are studying in Iran or receiving training in Libya. In May and June 1990 large Muslim rallies took place in the southern Thai provinces led by an Iran-linked Shia activist. A subsequent warrant for the activist's arrest led him and several followers to escape to Malaysia.

A number of political groups have sought to represent the cultural, economic and political aspirations of the Malay Muslim minority (Dulyakasem 1981: 91–8). Any such separatist groups are an embarrassment to Malaysia in that, on the one hand, Malaysia cannot be seen to support them, in order to maintain good relations with Thailand, and, on the other hand, cannot be seen to condemn them for fear of alienating sympathetic Kelantan Malays (Keyes 1987: 132).

Racial identification and insecurity combined with a heightened sense of regionalism were prominent issues in the October 1990 Malaysian General and State elections and clearly had important implications for the Thai–Malaysian border region. The elections were important in any event because for the first time since Independence in 1957 a viable opposition coalition (Semangat '46) was offering itself as an alternative government. In the event, although the incumbent Barisan Nasional (BN) coalition retained the necessary two-thirds majority with its 127 seats, there were some very significant electoral shifts, especially in the peripheral regions. This was particularly the case

DENNIS RUMLEY

in the border state of Kelantan and in Sabah. In the Kelantan State elections, for example, PAS (Parti Islam of Malaysia), one of whose central aims has been the creation of an Islamic state in Malaysia, won all 24 of the seats it contested, while BN representation fell from 29 seats in 1986 to none at all in 1990 (Table 7.2).

Table 7.2 Kelantan State election results 1986 and 1990

Party	1986		1990	
	Contested	Won	Contested	Won
Barisan Nasional	39	29	39	0
PAS	39	10	24	24
Semangat '46	0	0	14	14

Control by PAS of the Kelantan State government is potentially of great significance in the context of border-region relationships since, among other things, PAS is perceived by the Thais as in conflict both with the Malaysian federation and with Thailand itself. For example, the Thais believe that PAS has been involved in aiding Thai Muslim secessionists in the south despite Kuala Lumpur's official disapproval of Malaysian intervention in Thailand's internal affairs. This in turn feeds into Malaysian federal–state conflict primarily centred around the unusual situation of federal and state governments being of different political persuasions. At worst, this situation is seen by the federal government as the first step to secession; at best, it fosters intense mutual suspicion, especially in terms of joint federal–state development initiatives. One expression of the latter was the decision by Prime Minister Mahathir to call a meeting in November 1990 of all state Chief Ministers plus the Federal Cabinet to co-ordinate federal–state activities. However, the Chief Ministers of both Kelantan and Sabah were not invited.

Economic peripherality

The Thai–Malaysian border region is economically peripheral to its respective states by any measure (Rumley 1991c). This therefore introduces a further potential source of conflict into the region which interacts with the other dimensions. Overall, the border economy is

predominantly rural and agricultural and is dominated by rice culti-
vation (for example, in Pattani and in Kedah) and by the production of
rubber (for example, in Yala). In the five Thai border provinces, employment in the agricultural sector ranges from a low of 67 per cent in
Songkhla to a high of 77 per cent in Pattani (National Statistical Office
1981). It has been estimated that 80 per cent of Malay Muslims in
southern Thailand are involved in rice farming and rubber plantations,
all on relatively small average landholdings (Pitsuwan 1985: 19). In
northern Malaysia, on the other hand, Perak and Terengganu have had
a traditionally greater emphasis on mining and manufacturing
compared with most of the other border states which are predominantly
agricultural (Pryor 1978: 64).

In Malaysia, contributions to the Gross Domestic Product (GDP) of
the three most important industrial sectors (agriculture, manufacturing
and mining and wholesale and retail trade) show that Perak is the only
border state with an above-average contribution in all three sectors. Of
the remaining four states, in the main, all have significantly below-
average GDP contributions with the exception of Kedah for agriculture
(*Fifth Malaysian Plan* 1986: 174–5).

In Thailand the agricultural economy of the border provinces has
been in relative decline since World War II. This is a direct reflection of
the national shift in the contribution made by the various sectors to
economic growth in the country. In 1951–84, for example, the contribution to economic growth made by the agricultural sector declined
from 50.1 per cent to 19.9 per cent without any corresponding shift
from agricultural to non-agricultural occupations. The most rapid
growth, on the other hand, has been in the industrial sector (Keyes
1987: 154). In southern Thailand the rural less-educated population
faces serious economic difficulties. For example, rubber prices,
important to the local economy, have been variable and have even
fallen, while at the same time prices for staple food such as rice have
been rising.

In Malaysia as a whole there has been a well-defined communal
complexion to income disparities between 'rich' and 'poor' states, with
the Malays being heavily concentrated in the relatively poor states of the
border region with the possible exclusion of Perak. In 1985 the degree
of poverty in the border states ranged from 20.3 per cent in Perak to
39.2 per cent in Kelantan, compared with a total of 18.4 per cent for all
of peninsular Malaysia; furthermore, unemployment has been consistently above the national average in the five border states (*Fifth Malaysian Plan* 1986: 88, 170–1).

The interaction of geographical, socio-cultural and political peripherality with economic peripherality is expressed in at least two major types of economic conflicts in the Thai–Malaysian border region – via the Thai Muslim community on the one side and as a result of the outcome of the politics of Malaysian resource allocations on the other. Economic-development policies applied to similar groups on either side of the boundary clearly reflect different national goals. For some in the Thai Muslim community it is likely that a significant element in their conflict with the central government is as much a reflection of economic inequalities as it is about Islamic fundamentalism. Many southern Thai Muslims feel that they have little or no part in Thailand's economic boom. Furthermore, the Muslim community has an extremely limited role in southern business in any event, partly due to inadequate education.

In the border region of Malaysia, on the other hand, there is a set of interrelated conflicts arising out of the interaction between economic inequality, ethnicity, political peripherality and governmental resource allocations. This is especially evident in Kelantan where, in order to try to circumvent the PAS administration, the federal government has formed a new Federal Development Office which is directly under the control of the Prime Minister's Department. The principal aim of this is to oversee federal projects in Kelantan with field officers who are federal civil servants. In addition, the federal government's withdrawal of fertilizer subsidies is likely to have a significant regional impact on the next rice harvest and place a difficult financial burden on the Kelantan state government.

Malaysia's New Economic Policy (NEP), launched in 1971 not long after the 1969 race riots, expired in December 1990 and was replaced in June 1991 with a New Development Policy (NDP). One of the main aims of the NEP was a positive-discrimination policy of redistributing income and employment to the state's *bumiputra* ('sons of the soil', with reference primarily to the Malay population). Theoretically, such a policy was especially beneficial to the peripheral border states since they contain the largest proportions of Malays (Rumley 1991c). Although the original NEP goals will be largely retained, the emphasis in the NDP will be on 'growth with equity' which is mainly designed to alleviate any fears on the part of non-Malays over the new policy. However, an overriding aim will continue to be the improvement of the condition of *bumiputras*, especially in the business and manufacturing sectors. Although racial quotas will be de-emphasized they will not be phased out. As far as the states of the Malaysian border region are concerned, any dilution of the original NEP goals is likely to be detrimental and thus reinforce conflicts arising out of peripherality.

CONCLUSIONS

An understanding of the precise scope and nature of inherent border-region conflict is clearly the first step in moving towards government policies which emphasize co-operation. It has been argued here that such an understanding derives from a synthesis of the various approaches to border-region conflict within an overall centre–periphery framework.

There are a number of examples which can be cited of border regions in which former conflict has been replaced by policies which stress co-operation. Indeed, research into and support for cross-boundary developments has been one of the key questions in some long-term European research programmes (Galluser 1991). As far as the Thai–Malaysian border region is concerned, the thirty-fifth meeting of the Malaysia–Thailand General Border Committee (GBC) held in Kuala Lumpur in July 1991 concluded that the security situation had improved considerably since the cessation of armed struggle of the Communist Party of Malaya in December 1989. Both Thai and Malaysian delegations agreed that efforts should now focus on stabilizing the border region through socio-economic development. It was agreed that the Socio-Economic Development Committee of the GBC should now play a major role in identifying mutually beneficial joint economic projects. These could build on the positive experience arising from the earlier establishment of the Malaysia–Thailand Joint Authority which came into effect in January 1991 for joint exploration and exploitation of the continental shelf in the Gulf of Thailand for petroleum. The timing and relative success of moving from a context of conflict to an environment of co-operation remains to be seen.

NOTE

1   The author would like to express his sincere appreciation to the Indian Ocean Centre for Peace Studies, University of Western Australia and Curtin University of Technology, and especially to its Executive Director, Ken Macpherson, for continued support and encouragement without which the preparation of this chapter would not have been possible.

REFERENCES

Brundtland, G.H. (1990) *Our Common Future*, Melbourne: Oxford University Press.

Cady, J.F. (1974) *The History of Post-War Southeast Asia*, Athens, Ohio: Ohio University Press.

Carstens, S.A. (ed.) (1986) *Cultural Identity in Northern Peninsular Malaysia*, Athens, Ohio: Centre for International Studies.

Chai-anan, S. (1989) 'Thailand: A Stable Semi-Democracy', in L. Diamond, J.J. Linz and S.M. Lipset (eds), *Democracy in Developing Countries, vol. III, Asia*, Boulder: Rienner, 304–46.

Drummond, J. and Manson, A.H. (1991) 'The Evolution and Contemporary Significance of the Bophuthatswana-Botswana Border Landscape', in D. Rumley and J.V. Minghi (eds), *The Geography of Border Landscapes*, London: Routledge, 217–42.

Dulyakasem, U. (1981) 'A Study of Muslim-Malays in Southern Siam', Ph.D. dissertation, Stanford University.

FEER (Far Eastern Economic Review), 1 November 1990, 6 June 1991.

Fifth Malaysian Plan, 1986–1990 (1986), Kuala Lumpur: Government of Malaysia.

Galluser, W.A. (1991) 'Geographical Investigations in Boundary Areas of the Basle Region ("Regio")', in D. Rumley and J.V. Minghi (eds), *The Geography of Border Landscapes*, London: Routledge, 31–42.

Haemindra, N. (1976) 'The Problem of the Thai-Muslims in the Four Southern Provinces of Thailand (Part One)', *Journal of Southeast Asian Studies*, 7: 197–225.

Hall, D.G.E. (1981) *A History of Southeast Asia*, London: Macmillan.

Hurst, P. (1990) *Rainforest Politics: Ecological Destruction in Southeast Asia*, London: Zed Books.

Johnston, R.J. (1989) 'The State, Political Geography, and Geography', in R. Peet and N. Thrift (eds), *New Models in Geography*, London: Unwin Hyman, vol. I: 292–309.

Keyes, C.F. (1987) *Thailand: Buddhist Kingdom as Modern Nation-State*, Boulder: Westview.

Kliot, N. (1991) 'The Political Geography of Conflict and Peace – An Introduction', in N. Kliot and S. Waterman (eds), *The Political Geography of Conflict and Peace*, London: Belhaven Press, 1–17.

Kliot, N. and Waterman, S. (eds) (1991) *The Political Geography of Conflict and Peace*, London: Belhaven Press.

Knight, D.B. (1983) 'The Dilemma of Nations in a Rigid State Structured World', in N. Kliot and S. Waterman (eds), *Pluralism and Political Geography: People, Territory and State*, London: Croom Helm, 114–37.

Lamb, A. (1968) *Asian Frontiers: Studies in a Continuing Problem*, London: Pall Mall Press.

Lane, J.E. and Ersson, S. (1989) 'Unpacking the Political Development Concept', *Political Geography Quarterly*, 8: 123–44.

Lee, Y.L. (1982) *Southeast Asia: Essays in Political Geography*, Singapore: Singapore University Press.

Lim, J.J. (1984) *Territorial Power Domains, Southeast Asia and China,* Singapore: Institute of Southeast Asian Studies.
Mandel, R. (1980) 'Roots of the Modern Inter-State Border Dispute', *Journal of Conflict Resolution,* 24: 427–54.
Minghi, J.V. (1991) 'From Conflict to Harmony in Border Landscapes', in D. Rumley and J.V. Minghi (eds), *The Geography of Border Landscapes,* London: Routledge, 15–30.
Nation, 5 June 1991.
National Statistical Office (1981) *Population and Housing Census 1980,* Bangkok: Office of the Prime Minister.
Nyerere, J.K. (1990) *The Challenge to the South: The Report of the South Commission,* Oxford: Oxford University Press.
O'Tuathail, G. (1987) 'Beyond Empiricist Political Geography: A Comment on Van Der Wusten and O'Loughlin', *Professional Geographer,* 39: 196–7.
Pitsuwan, S. (1985) *Islam and Malay Nationalism: A Case Study of the Malay-Muslims of Southern Thailand,* Bangkok: Thai Khadi Research Institute.
Prescott, J.R.V. (1990) *Political Frontiers and Boundaries,* London: Unwin Hyman.
Pryor, R.J. (1978) 'Internal Migrants in Peninsula Malaysia', *Journal of Tropical Geography,* 46: 61–75.
Rachagan, S.S. and Dorall, R.F. (1976) 'Rivers as International Boundaries: The Case of the Sungei Golok, Malaysia–Thailand', *Journal of Tropical Geography,* 42: 47–58.
Rumley, D. (1991a) 'Politics and Development: A Report on the IGU Commission on the World Political Map Conference in New Delhi, India, 1–4 November 1990', *Political Geography Quarterly,* 10: 317–20.
—— (1991b) 'The Political Organisation of Space: A Reformist Conception', *Australian Geographical Studies,* 29/2, 329–36.
—— (1991c) 'Society, State and Peripherality: The Case of the Thai–Malaysian Border Zone', in D. Rumley and J.V. Minghi (eds), *The Geography of Border Landscapes,* London: Routledge, 129–51.
Rumley, D. and Minghi, J.V. (eds) (1991a) *The Geography of Border Landscapes,* London: Routledge.
—— (1991b) 'Introduction: The Border Landscape Concept', in D. Rumley and J.V. Minghi (eds), *The Geography of Border Landscapes,* London: Routledge, 4–6.
Shafruddin, B.H. (1987) *The Federal Factor in the Government and Politics of Peninsular Malaysia,* Singapore: Oxford University Press.
Spencer, J.E. and Thomas, W.L. (1971) *Asia, East By South: A Cultural Geography,* New York: Wiley.
Taylor, Peter J. (1991) 'If Cold War Is the Problem, Is Hot Peace the Solution?', in N. Kliot and S. Waterman (eds), *The Political Geography of Conflict and Peace,* London: Belhaven Press, 78–92.
Taylor, P.J. and Johnston, R.J. (1979) *Geography of Elections,* Harmondsworth: Penguin, 112.
Tinker, H. (1990) 'The Race Factor in International Politics', in P. Smoker, R. Davies and B. Munske (eds), *A Reader in Peace Studies,* Oxford: Pergamon, 43–50.
Van Der Wusten, H. and O'Loughlin, J. (1986) 'Claiming New Territory for a

Stable Peace: How Geography Can Contribute', *Professional Geographer*, 38: 18–28.

Wellhofer, E.S. (1989) 'Core and Periphery: Territorial Dimensions in Politics', *Urban Studies*, 26: 340–55.

Wheatley, P. (1961) *The Golden Khersonese*, Kuala Lumpur: University of Malaya Press.

Wijeyewardene, G. (ed.) (1990) *Ethnic Groups Across National Boundaries in Mainland Southeast Asia*, Singapore: Institute of Southeast Asian Studies.

Yiftachel, O., Carmon, N. and Rumley, D. (1991) 'The Political Geography of Minority Control in Israel and Malaysia', in N. Kliot and S. Waterman (eds), *The Political Geography of Conflict and Peace*, London: Belhaven Press, 184–200.

Zakaria, H. A. (1989) 'Malaysia: Quasi Democracy in a Divided Society', in L. Diamond, J.J. Linz and S.M. Lipset (eds), *Democracy in Developing Countries, vol. III, Asia*, Boulder: Rienner, 304–46.

8

THE HONG KONG–CHINA BORDER

Under whose management?

Patricia Goodstadt

INTRODUCTION

Until recently the Hong Kong–China border has performed two important functions: it has served as a unique bridge between China and the outside world, and it has acted as a barrier between Hong Kong and the effects of mainland policies and events. Even during China's most isolationist years, the border remained open to some degree of trade and contact between Hong Kong and China; this permeability has been in the interests of both China and Hong Kong. Insulation from Chinese events has been far from absolute, but the border has shielded Hong Kong from China's political, economic and legal systems, and has fostered in the territory a stability markedly absent from modern China. Since China's economic reform programme began in 1978, Hong Kong has expanded vastly its role as a link between China and capitalist countries. Over the past fourteen years the nature of the boundary relationship has begun to change.

Two of the most significant changes since 1978 in the boundary relations between Hong Kong and China are the closure of the boundary to illegal immigration into Hong Kong and the opening of the Chinese border to investment and exports from Hong Kong. This chapter discusses various aspects and implications of these two changes in the context of Hong Kong's future reintegration in China.

BACKGROUND

The present territory of Hong Kong was ceded to the British in a series of three treaties[1] following China's defeat in wars[2] (Hsu 1982: 189–273, 406–29) with Western nations in the nineteenth century. Since the

People's Republic of China views the treaties signed during this period as 'unequal treaties' imposed at gunpoint, it denies their validity and has held that the British rule in Hong Kong is illegitimate (Wesley-Smith 1980). The boundary imposed by these treaties has, however, created a Chinese city (H. Lethbridge 1984: 52, 55; Lau and Kuan 1988, ch. 1) insulated from the effects of policies pursued in mainland China. Without the protection of this barrier, Hong Kong could not have experienced its exceptional economic development (Chen 1984: 1–13) and a social stability (Miners 1975; Lau 1982; and Miners 1986: 31–42) markedly contrasting with the turmoil prevalent in twentieth-century China.[3] A sizeable segment[4] of Hong Kong's population has made a deliberate choice to live on the Hong Kong side of the boundary by migrating to the colony from China. Hong Kong's attraction was not solely economic; protection from the vagaries of Chinese politics was at least as important. As a Chinese member of Hong Kong's Legislative Council remarked, 'Hong Kong is the lifeboat, China is the sea' (Hoadley 1970: 211). Most of these migrants came from adjoining Guangdong Province, to which Hong Kong belongs linguistically (Moser 1985) and culturally. The majority of Hong Kong's population are Cantonese and have family ties across the boundary. Despite the barrier of an international boundary, the People's Republic of China has permitted sufficient permeability at the boundary for these ties to be maintained.

The Last Post for British colonial rule in Hong Kong was sounded in 1984, when Britain promised to hand Hong Kong back to China in 1997. Under the terms of the Sino-British *Joint Declaration* (*Joint Declaration of the Government of the United Kingdom of Great Britain and Northern Ireland and the Government of the People's Republic of China on the Future of Hong Kong* 1984), Hong Kong will become a Special Administrative Region of the People's Republic of China in 1997. Its capitalist system will be preserved under the 'one country, two systems' formula proposed by the Chinese. However, China has not spelt out fully how this concept is to function in practice.

After 1997 the international boundary between Hong Kong and China will no longer exist, but a Hong Kong–China boundary will remain, as an internal administrative border. This border is fundamental to the operation of the 'one country, two systems' formula. The boundary will need to function simultaneously as a dividing and unifying device, as it will be the means by which capitalism and communism are kept separate and the means by which Hong Kong is integrated into China. By the study of China's policies towards this boundary, a clearer understanding can be gained of how China expects Hong Kong to function in 1997.

THE BOUNDARY BETWEEN HONG KONG AND CHINA

The political border operated from 1949 until 1978 to mark a clear boundary for the People's Republic of China, a socialist and ideologically directed society,[5] from a capitalist, imperialist world. A fundamental tenet of Maoist ideology was that the capitalist, imperialist world was hostile and that contacts with it must be subject to close regulation.

The Central People's Government exerted almost total control over the border with Hong Kong and over the conditions under which contacts would be permitted between Hong Kong and the People's Republic of China.[6] The Central People's Government exercised legal and institutional control over the movement of money, goods and people across the border. It also had ultimate control over most economic transactions and operations (Ma 1990). Within the Chinese borders, even social activity (including dating and marriage) came under Party or government control. Cross-border flows therefore reflected two factors: first, Chinese government policy decisions about the type and degree of contacts which would be in the interests of the People's Republic of China; and second, the conditions which Hong Kong parties would encounter on the Chinese side of the boundary.

Despite the isolationist policies practised throughout much of the history of the People's Republic of China, the Hong Kong–China boundary had never been completely sealed off. On the Hong Kong side, there have been very few restrictions on the flow of people, goods or money in or out of Hong Kong's borders (Jao 1983; Rabushka 1979; Youngson 1983). Before 1979 there were two important cross-border flows from China to Hong Kong: exports and immigrants. Hong Kong has received a consistent stream of exports (mainly foodstuffs, water, raw materials and low-cost consumer items) from China (Hsueh and Woo 1981: 1–4, 9–11; Chau 1983; Schiffer 1983: 8–9). Despite its communist ideology, the People's Republic of China showed itself willing and able to behave competitively in the supply of goods to Hong Kong (Jao 1983: 23; Chau 1983: 191–8) throughout its history. It has successfully maintained a fairly constant share of Hong Kong's total imports since 1949 (Figure 8.1), despite competition from south-east Asian suppliers, particularly since the 1970s. Since 1949 there have

133

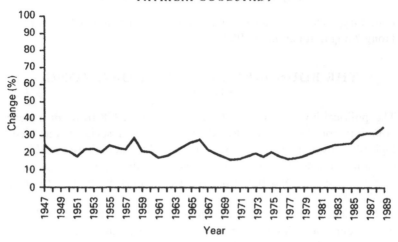

Figure 8.1 Imports from China as a percentage of total Hong Kong imports
Source: Derived from data in Census and Statistics Department, HK (1990).

been only five years in which China's exports to Hong Kong did not show an increase over the previous year (Figure 8.2); China did not allow its exports to Hong Kong to be affected by internal political or ideological campaigns, except in extreme circumstances. These exports, essential to Hong Kong's existence and economy (Chau 1983; Chen 1984: 23), have also been an important source of foreign exchange for China.

Until 1980 Hong Kong's boundary was semi-open to illegal immigration from China into Hong Kong.[7] Legal and illegal immigration from China have been an essential source of capital, know-how[8] and labour for Hong Kong. The number of illegal immigrants entering Hong Kong ranged from several thousands to several hundred thousands each year (Table 8.1), depending on conditions inside China. Until October 1980 any Chinese illegal immigrant who reached the urban areas of Hong Kong was allowed to remain in the territory. Up to that date, Hong Kong employers could rely on a supply of labour adequate to meet requirements created by an expanding economy.

On the Chinese side, China's borders were largely closed to imports until it began to open up to the outside world in the 1970s. Hong Kong's exports and re-exports to China remained at low levels until after 1979, both as a proportion of the Hong Kong total (Figures 8.3a and 8.4a) and in terms of their actual values (Figures 8.3b and 8.4b). However, there have been certain important economic flows from Hong

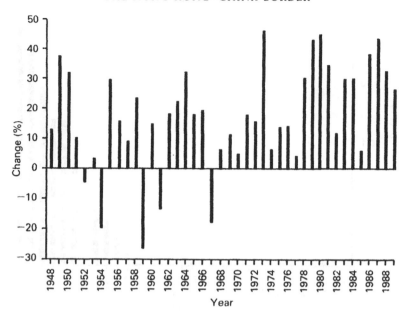

Figure 8.2 Annual percentage change in China's exports to Hong Kong
Source: Derived from data in Census and Statistics Department, HK (1990).

Table 8.1 Illegal immigration from China into Hong Kong (selected years)

Year	Arrested	Evaded arrest (estimates)
1977	1,965	6,500
1978	8,205	27,400
1979	89,942	11,000
1980	80,320	68,501
1983	4,671	–

Source: Royal Hong Kong Police Force (1978–81; 1984), *Annual Report.*

Kong to China. Hong Kong is southern China's main source of remittances,[9] which are sent back by Hong Kong residents to relatives in China (one of the benefits to China of permitting family ties to be maintained across the border). Hong Kong is also one of China's most important sources of foreign exchange. It is estimated that Hong Kong

135

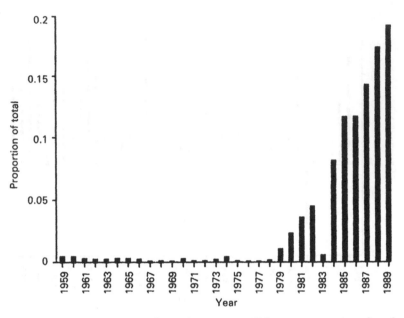

Figure 8.3a Hong Kong's domestic exports to China as a proportion of total
domestic exports
Source: Derived from data in Census and Statistics Department, HK (1990).

contributes around one-third of China's gross foreign-exchange earnings (Jao 1983: 38–42; Economic Services Department, Hong Kong
Government, unpublished figures 1983, 1987). The links with Hong
Kong also enabled China to obtain (limited) supplies of otherwise
unavailable goods during the Korean War embargo, earn hard currency
and maintain some unofficial economic and political contacts with the
outside world. Hong Kong's almost completely open borders thus
enabled it to act as a 'window' for China at a time when China's boundaries were almost entirely closed to Western capitalist nations.

Changes in the border relationship

The border relationship changed markedly in 1978–9. China's boundaries became much more open to the outside world, particularly with
Hong Kong. The Chinese authorities now see trade, external investment
and co-operation with industrialized nations as indispensable to
national economic development; a new 'open door' strategy for China's
economic development makes a deliberate effort to woo investment

136

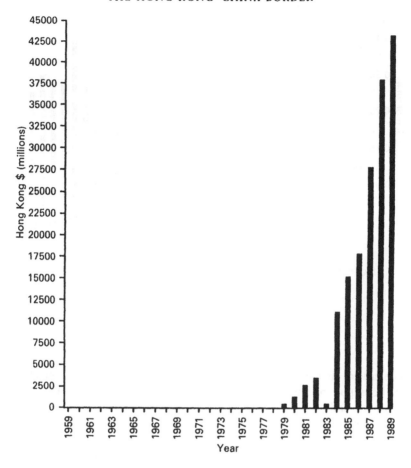

Figure 8.3b Hong Kong's domestic exports to China
Source: Derived from data in Census and Statistics Department, HK (1990).

from abroad.[10] Hong Kong has been targeted as a particularly promising source of investment, technology and expertise (*South China Morning Post*, 30 March 1979; Feng 1989; Fu 1989) and its neighbouring province Guangdong has been given special autonomy to establish flexible systems and incentives for external trading and investment partners. Three out of the four Special Economic Zones set up to attract external investment are also in Guangdong (Su 1985).[11] The choice of location anticipated an important role for Hong Kong in supplying goods, know-how and funds for China's modernization.

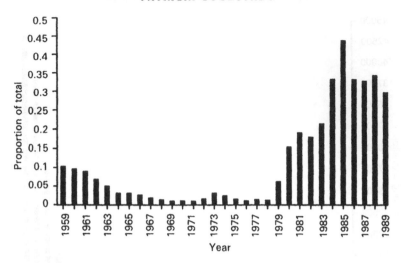

Figure 8.4a Re-exports to China as a proportion of total Hong Kong
re-exports
Source: Derived from data in Census and Statistics Department, HK (1990).

IMMIGRATION

The crucial change in the relationship on the Chinese side was the end of the export of labour to Hong Kong. On the Hong Kong side, the desire to end the free importation of labour in the form of illegal immigrants was inspired by social anxieties and worries over public security (Royal Hong Kong Police Force 1979: 15). For China, the exodus to Hong Kong had been violating its long-standing policy of controlling movement of its population between rural and urban areas and among urban centres.[12] An illegal emigration which the Chinese authorities were seemingly unable to halt was potentially embarrassing for the Chinese government. Agreements signed between the Hong Kong Government and the Guangdong provincial authorities in the 1970s and 1980s created a system which allowed the Hong Kong Government to send back to China all unlawful entrants into the territory.

The simple solution to the problem of large-scale illegal immigration would have been for the Hong Kong Government to deport all the unlawful entrants promptly, permanently and with minimal formality. The obstacle to this approach was the Central People's Government's view on sovereignty (Lane 1990; Dicks 1983). The Central People's

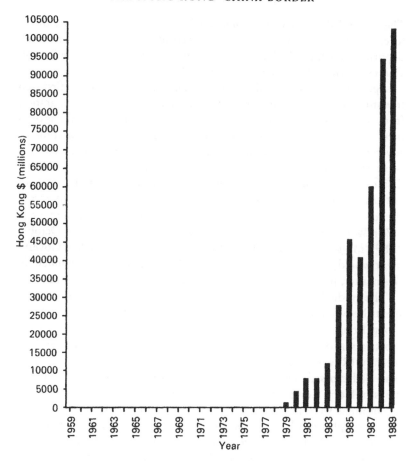

Figure 8.4b Hong Kong's re-exports to China
Source: Derived from data in Census and Statistics Department, HK (1990).

Government has taken the consistent stand that the United Kingdom was unlawfully in possession of the territory of Hong Kong, having acquired it through 'unequal treaties' imposed on the Ch'ing dynasty of China in the last century:

> The questions of Hong Kong and Macao belong to the category of questions resulting from the series of unequal treaties which the imperialists imposed on China. Hong Kong and Macao are part of Chinese territory occupied by the British and Portuguese authorities.[13]

139

The Central People's Government could not officially condone, therefore, any measures by the British that permitted China's citizens to be deported or expelled (depending on the legal process involved) from Chinese territory by the colonial power.[14]

Agreements were needed with the Chinese authorities as the first step towards the enforcement of full and effective control over immigration from the People's Republic of China. These agreements[15] would have to meet Central People's Government sensitivities on the question of sovereignty over Hong Kong. Since the Chinese government did not want Hong Kong to become a haven for those out of favour with the Chinese Communist Party leadership, nor did it want a centre of criticism on what it regarded as Chinese soil,[16] it was willing to accept the return of all illegal immigrants, providing no exceptions were made (*Wen Wei Pao*, 27 September 1989; *Hong Kong Times*, 17 October 1989; *Ming Pao*, 17 October 1989).[17] The Central People's Government would, therefore, co-operate with the Hong Kong Government in this matter, provided the right diplomatic formula could be found.

The gradual establishment of a system of formal and informal agreements between the two governments slowed the flow of population into Hong Kong. From 1974 the Chinese authorities agreed to accept the repatriation of mainland Chinese caught trying to enter the territory illegally (Royal Hong Kong Police Force 1975). This included those who had crossed the border, but had been caught outside the urban areas of Hong Kong, although illegal immigrants who had reached the urban areas were permitted to stay. In 1980 the Hong Kong Government brought to a halt for the first time since 1949 all unauthorized immigration. All those who left the People's Republic of China unlawfully were returned, regardless of where they had been discovered inside Hong Kong, with the active co-operation of the People's Republic of China in the process (notably, accepting the delivery of persons at the boundary from the custody of the Royal Hong Kong Police Force). The Chinese authorities also began to co-operate actively with the Hong Kong Government in stricter border controls to prevent illegal population movements (Royal Hong Kong Police Force 1983: 7; *Hong Kong Hansard* 1983: 31).

The numbers and status of those arriving legally[18] were also strictly regulated, thanks to controls imposed by the Central People's Government. Before 1979 there had been an informal agreement that Guangdong residents from China would be permitted to enter Hong Kong, subject to a limit of fifty each day (Immigration Department, *Annual Reports*, for various years, in particular, the Report for 1961–2:

9). Until the 1960s that limit was never under pressure, but by 1967 Hong Kong immigration authorities had difficulties in enforcing any numerical limit.[19] By 1978 a daily average of 185 legal migrants with Chinese exit permits were entering Hong Kong. The number of Chinese applications for Hong Kong visas was such that 'even the physical process of recording them had become overwhelming'.[20] In 1979 the Hong Kong Government stopped issuing visas to Chinese citizens applying from addresses in China for permission to enter Hong Kong. An understanding was reached with China whereby 150 Chinese residents would be admitted to Hong Kong each day (Hong Kong Government, Confidential Information Note, February 1990. The Hong Kong Government would allow persons to enter the territory provided that they held a valid exit permit from China for either a fixed period ('two-way permit') or permanent stay ('one-way permit').

The permits were issued by the Central People's Government without consultation with the Hong Kong Government. Possession of a valid permit was, to all intents and purposes, a guaranteed right to enter a British Crown Colony issued exclusively by the Chinese Government. The Hong Kong Government gained Chinese co-operation in minimizing the number of entrants into Hong Kong, but control of entry itself had passed to the Central People's Government.

The closure of the Hong Kong–China boundary to illegal immigration has had several effects favourable to China. The first was that China came to determine, in practice, who had the right to enter Hong Kong from China, in return for co-operation with the British administration in limiting the flow of people to Hong Kong.

In effect, the Central People's Government was taking the line that co-operation with the Hong Kong Government was based on British acceptance of China's right to determine who should be returned to Chinese jurisdiction. The Hong Kong Government should have administrative rights in operating the scheme, but the right to select individuals to remain in the territory contrary to Central People's Government policy should be abandoned. The Central People's Government wished to be able to treat Hong Kong as it would deal with a provincial or city administration within the People's Republic of China. The Chinese government apparently did not spell out its aims in this way, either publicly or privately. The British side were perhaps unaware of the full implications of the Central People's Government's standpoint.

The blueprint for future immigration policies

The de facto situation which emerged in the 1970s and early 1980s with the various immigration agreements between the Hong Kong and the Chinese authorities has been formalized in the Joint Declaration (1984) and the Basic Law. The Central People's Government Annex II to the Joint Declaration makes a clear distinction between post-1997 entry into Hong Kong of 'persons from other parts of China' and foreign visitors. In the former case, their entry 'shall continue to be regulated in accordance with the present [1984] practice'. The Central People's Government could argue that on the relevant date decisions on who could move from the People's Republic of China to Hong Kong for permanent residence were the Central People's Government's right. The next clause in Annex II gives Hong Kong the right to 'apply immigration controls on entry, stay and departure ... by persons from foreign states and regions' (which was existing practice in 1984). The Joint Declaration is laying down Central People's Government policy that the political boundary between Guangdong province and Hong Kong is different (paradoxically, to be subject to tighter Central People's Government oversight) from the boundary between Hong Kong and the rest of the world. Hong Kong will have control over the boundary between the Special Administrative Region and the rest of the world, but not over the boundary between Guangdong and Hong Kong. This distinction suggests that the Chinese government views the boundary as an important means of ensuring that Hong Kong will function as part of the Chinese state system and that its separate economic system will not confer administrative independence (as opposed to the 'autonomy' promised to Hong Kong under the Joint Declaration).

The conclusion here is that the Central People's Government had what appears to be a sharply defined policy towards the boundary in immigration matters. This policy is apparently aimed at extending China's control over movements by its nationals to Hong Kong and the curtailment of the Hong Kong Government's standing in this sector.

IMPLICATIONS FOR THE HONG KONG ECONOMY

The change in the permeability of the boundary to population flows from China is associated with changes in its economic functions. The control over emigration was to have long-term implications for the Hong Kong economy which, once again, the Hong Kong Government did not canvass in advance but which fitted in reality with the plans of the Central People's Government.

A key aim of the 'open door' policy was to attract external investment into China – from Hong Kong in particular. China hoped to use outside capital to create export capacity and to generate hard-currency receipts ('Resolution of the Standing Committee of the NPC Approving Regulations on Special Economic Zones in Guangdong', 26 August 1960, Articles 9 and 17; 'Law of the People's Republic of China on Foreign Capital Enterprises', 12 April 1986, Article 3). Despite implementing a range of special incentives to attract foreign capital, the People's Republic of China did not offer one incentive of great interest to the outside business community – the opportunity of establishing a market among the billion inhabitants of China. Although China's boundaries were opened to allow an inflow of foreigners and of external capital, an economic barrier remained which kept the national market largely impenetrable.

The incentives offered (cheap land and labour, tax relief, import concessions and some legal protection from the more onerous features of socialism), however, were not unique in the Asian Region. As 'pull' factors, their value to investors with an international portfolio was comparatively low. The inflow of capital into China has been limited, especially if the Hong Kong contribution is excluded (Table 8.2). The ability to generate an inflow from Hong Kong which is so disproportionate to the territory's ranking among the world's economies needs further discussion.

The 'pull' factors mentioned above have been attractive to Hong Kong investors. Land prices and labour costs are especially important (Chai 1983). Hong Kong has land values among the highest in the world. In the People's Republic of China, a commercial land market is relatively new. Article 10 of the 1982 *Constitution* states that:

Land in the cities is owned by the state.

Land in the rural and suburban areas is owned by collectives except for those portions which belong to the state in accordance with the law.

No organization or individual may appropriate, buy, sell or lease land or otherwise engage in the transfer of land by unlawful means.

The People's Republic of China constitution was amended only in 1988 to allow for private ownership of land. Firms were thus able to negotiate for sites (more cheaply than in Hong Kong) outside the operations of a land market. Labour costs and availability were an important consideration. The official data on the Hong Kong labour market are set out in

143

Table 8.2 China's total realized foreign investment and Hong Kong's
contribution (US$ 000s)

Year	Chinese total/ Hong Kong's share	Total	External loans	Direct foreign investment	Others*
1979–83	National total	2,178,496	1,433,240	745,256	–
	Hong Kong	440,511	8,592	431,919	–
	(% of total)	(20.2)	(0.6)	(57.9)	
1984	National total	270,452	128,567	141,885	–
	Hong Kong	75,381	628	74,753	–
	(% of total)	(27.9)	(0.5)	(52.6)	
1985	National total	446,211	250,596	195,615	–
	Hong Kong	101,637	6,069	95,568	–
	(% of total)	(22.8)	(2.4)	(48.9)	
1986	National total	725,830	501,457	187,489	36,884
	Hong Kong	157,285	24,424	113,237	19,634
	(% of total)	(21.7)	(4.9)	(60.7)	(53.2)
1987	National total	845,156	580,459	231,354	33,308
	Hong Kong	207,227	28,219	158,947	20,762
	(% of total)	(24.5)	(4.9)	(68.7)	(63.2)
1988	National total	1,022,639	648,673	319,368	54,598
	Hong Kong	287,425	57,965	206,760	32,700
	(% of total)	(28.1)	(0.9)	(64.7)	(59.9)

Source: Derived from Almanac of China's Foreign Economic Relations and Trade, relevant years.
Note: 'Hong Kong' refers to investment from Hong Kong. Figures for 1979 to 1986 include capital from Macao (a negligible amount); 1987–8 figures are for Hong Kong alone.
* Processing and assembly.

Table 8.3, and Table 8.4 shows the difference in wage rates across the border. The tax and import concessions appear of little relevance to Hong Kong firms since the territory is a free port with a tax regime lower than other comparable economies (Hung 1984: 181–2). In themselves, however, these factors were not dramatically more attractive to Hong Kong than other business centres in south-east Asia.

Furthermore, Hong Kong firms and investors had a major disincentive not shared by businesses from foreign countries. The bulk of Hong Kong's adult population took up residence in the territory out of a desire to avoid life under the Chinese Communist Party. They are the least likely to be easily induced by the Central People's Government to

Table 8.3 Hong Kong labour-market data (1980–8)

Year	GDP change per annum (%)	Payroll/person index (June 1980 = 100)	Labour-force participation rates (%)	Median weekly hours of work	Vacancy rate (%)	Unemployment rate (%)	Labour-force (000s)
1980		114.8	62.8	46	3.5	3.8	2323.4
1981	+9.5	121.3	63.1	47	3.9	3.9	2489.5
1982	+3.0	122.5	63.9	47	2.4	3.6	2498.1
1983	+6.5	125.1	64.5	46	2.8	4.5	2540.5
1984	+9.5	133.2	65.5	46	2.6	3.9	2606.2
1985	−0.1	142.7	64.8	47	1.8	3.2	2626.9
1986	+11.9	157.0	65.1	47	2.4	2.8	2701.5
1987	+13.0	159.1	64.8	47	3.7	1.7	2736.0
1988	+7.4	167.4	64.5	47	4.6	1.4	2778.6

Source: Census and Statistics Department, HK (1990).

Notes: 1 The Gross Domestic Product growth-rate is calculated at constant (1980) market prices.

2 Payroll per person refers to December of each year. The payroll per person indexes are estimated by weighting the payroll index of the sectors covered by the Survey of Employment, Vacancies and Payroll (SEVP) by persons engaged. The SEVP does not cover self-employed persons, the public sector and the construction sector.

3 Labour-force figures refer to the third quarter of the General Household Survey (GHS).

4 Median hours of work refer to third quarter of the GHS.

5 Vacancy rate is estimated by vacancies as a percentage of the sum of vacancies and persons engaged in the relevant quarter of the SEVP.

6 Unemployment rate is the percentage of the unemployed over the labour-force. Figures are averages of the quarterly GHS.

Table 8.4 Cross-border differences in the monthly wage for an
industrial worker

Area	US$
Hong Kong	412.0
Shenzhen	75.0
Guangdong	52.2
Beijing	46.9
Shanghai	50.3
Jiangsu	41.5
China (national total)	39.9

Sources: Figures for Hong Kong and Shenzhen are from Heng Seng Bank (1990) and are
for 1989; the remainder have been computed from China Statistical Yearbook 1990, and
refer to 1988.

put either their funds or their persons at risk on China's side of the
border. Hong Kong investors operating inside China also are at a
greater political risk than their foreign counterparts. Unlike firms and
individuals whose foreign status is recognized by the Central People's
Government and who are thus entitled to claim consular protection, the
ethnic Chinese residents of Hong Kong are not deemed by the Central
People's Government to be aliens with consular rights[21] (Fang 1986: 55;
Li 1986: 42–3).

Nevertheless, Hong Kong firms have been the predominant source of
investment capital both for the People's Republic of China in toto and
for Guangdong Province. The explanation is to be found in a 'push'
factor: the territory's shortage of labour became so acute that investment in Guangdong was seen as the most practical solution to the
inability of available supplies to meet the needs of an economy which
was growing rapidly.[22]

The immigration controls imposed in 1979–80 were of major
importance in creating this 'push' factor. They altered drastically the
nature of the territory's labour market. Up to 1980 Hong Kong
employers had been able to rely on finding adequate numbers of
additional workers to meet the requirements created by increased orders
as the economy expanded. Immigration from China provided an
abundant labour supply during the late 1940s and 1950s. In the 1960s
and early 1970s Hong Kong's post-war baby boom (accentuated by the
influx of Chinese immigrants of child-bearing age) and increased
labour-force participation rates among the female population continued
the supply of labour. After 1976 immigration from China[23] once again

became of vital importance, providing a valuable input at a time of rapid economic expansion in Hong Kong (Chen 1984: 13–16). These recruits to the labour force, most of whom were illegal immigrants, had several characteristics valuable to employers. Illegal crossing of the boundary involved a physically arduous and potentially dangerous exercise in evading heavy patrols by troops and police on both sides of the border. Crossing over sea could involve a swim across the shark-infested Mirs Bay. Once on Hong Kong territory the illegal immigrants had to make their way to the urban areas undetected. As a result, these newcomers tended to be fit, young and highly motivated.

The situation was reversed in the 1980s when illegal immigration virtually halted. The annual average increase in the work-force fell to 2 per cent compared with 5 per cent in the 1970s (Census and Statistics Department data). Table 8.5 shows the decreased rate of population growth in Hong Kong during the 1980s after the new immigration laws came into effect; Table 8.6 shows the drastic decrease in illegal immigrants from China after October 1980.

The ability of employers to tap other potential sources of labour in Hong Kong was limited by social developments brought about by increasing prosperity. The most important was the implementation of a universal, free secondary education linked to expansion of tertiary and vocational educational opportunities. Labour-force participation rates for the 16–19 age group fell from 40.05 per cent in 1978 to 29.95 per cent in 1988 (Labour Force Surveys, 1978; General Household Surveys, 1988; Census and Statistics Department, 1990). More generous social-security provisions permitted earlier retirement. Table 8.7 shows the

Table 8.5 Hong Kong population growth-rates

Year	Population growth-rate (%)	Crude birth-rate (/1000 persons)	Rate of natural increase (/1000 persons)
1978	1.8	17.3	12.2
1983	1.5	15.6	10.6
1984	1.1	14.4	9.6
1985	1.0	14.0	9.3
1986	1.4	13.0	8.3
1987	1.5	12.5	7.7
1988	1.2	13.3	8.4

Source: Census and Statistics Department, HK (1990: 1).

Table 8.6 The effect on illegal immigration into Hong Kong of the
October 1980 boundary closure

Year	Arrested	Evaded arrest (estimate)
1978	8,205	27,400
1979	89,942	11,000
1980 (total)	80,320	68,501
January–October	76,602	67,880
October–December	3,718	621
1983	4,671	–

Source: Royal Hong Kong Police Force, Annual Review 1978, 1979, 1980, 1981, 1984.

changes in labour-force participation rates for the relevant age groups. Table 8.3 presents plain evidence of a chronically tight labour market in which the shortage was aggravated over time even when Gross Domestic Product (GDP) growth-rates slackened.

The importance of labour as a factor of production markedly increased after 1980. Table 8.8 shows the relative shares in GDP annual growth attributable to capital and to labour for 1974–87. For the period as a whole, increase in GDP attributable to a marginal increase in the labour-force was 2.04 times (that is, elasticity of labour-force : elasticity of investment = 0.98/0.48) as large as the increased GDP attributable to a proportionate increase in investment; however, for the period 1980–7, after the importation of labour was ended, the additional GDP attributable to a change in the labour-force compared to a proportionate

Table 8.7 Labour-force participation rates (aged 55 and above) by sex and age group

Age	Sex	Labour-force participation rate in each age group		Change 1978–89 (%)
		1978	1988	
55–64	Male	74.4	69.0	−7
	Female	32.1	25.7	−19.9
65 and over	Male	33.8	23.9	−29.3
	Female	12.1	8.6	−28.9

Source: Derived from Census and Statistics Department, HK (1990).

Table 8.8   The relative shares in Hong Kong GDP growth attributable to
capital and to labour (1974–87)

	Elasticity of investment	Elasticity of labour-force
1974–9		
Total economy	0.48	0.98
1980–7		
Total economy	0.39	0.22
Manufacturing	0.50	−0.44
Non-manufacturing	0.33	1.7

Source: Unpublished Hong Kong Government data.

change in investment was 5.69 times as high (that is, 2.22/0.39). A large increase occurred in the importance of labour as a factor of production compared with investment after the end of labour imports from China. The magnitude of the increase suggests that immigration restrictions created a very strong 'push' factor for manufacturers' locational decisions. The difference between the labour elasticities for the manufacturing and non-manufacturing sectors is partly explained by the ability of Hong Kong manufacturers to tap China's labour supply through relocation of production. In general, China's economic environment is unattractive to tertiary and quaternary industry and China has not made a great effort to attract this sector of the economy.

Various scholars have noted that labour supply was an important factor in explaining the relocation of Hong Kong's productive capacity into the People's Republic of China. This chapter has shown that there was a clear tightening of Hong Kong's labour market after 1980 and that labour became much more important as a factor of production. These findings are consistent with empirical studies (notably Chen 1981, 1983a and 1983b) which have shown that the labour shortage is the most important reason in manufacturers' decisions to expand production across the boundary in China. The links between China's co-operation with Hong Kong on immigration and the northward shift of Hong Kong manufacturing have not, however, been hitherto discussed at any length. This chapter suggests that the Chinese decision to work together with the Hong Kong Government in controlling population movements across the boundary was a crucial factor in explaining why so many Hong Kong investors set up production plants in China.

ADAPTIVE STRATEGIES

The end to People's Republic of China labour exports in the form of illegal immigration from 1980 left Hong Kong with several strategies it could follow: to improve management of labour; to switch to less-labour-intensive methods of production; to substitute capital for labour; or to relocate production to somewhere with more abundant labour. The possibilities for the first three strategies are limited in Hong Kong. Hong Kong's industries are already operating efficiently, the Hong Kong economy is relatively sophisticated and the scope for replacing men with machines is relatively limited. The relationship between changes in productivity and wages in Hong Kong is shown in Table 8.9. The close link between the changes in productivity and changes in wages for the economy overall (E = 0.90) suggests that the Hong Kong labour market is relatively rational and efficient. Switching to less-labour-intensive methods of production is not easily accomplished within the short term (see Peat Marwick Management Consultants Ltd 1989). To shift manufacturing overseas also means various costs and constraints (Heng Seng Bank 1990).

On the other side of the equation, the factor to be considered is the strength of the 'pull' factor to shift to Guangdong Province. The crucial consideration here is the degree of compatibility between the business practices of a capitalist-style Hong Kong enterprise and those which would be permitted within the socialist-style environment of the People's Republic of China.

Cross-border economic incompatibility is greatest in the sectors of financial transactions and marketing. Hong Kong firms moving part of their activities to the People's Republic of China face the problem of minimum compatibility between their total freedom of financial transactions (Jao 1980: 177–9; Rabushka 1990: 336) in Hong Kong and their lack of such freedom in the People's Republic of China.[24] No matter what the other benefits, a Hong Kong firm must retain outside

Table 8.9 Earnings as a function of productivity in Hong Kong 1980–7

	R^2	Statistical significance (%)	Elasticity
Total economy	0.9200	1	0.90

Source: Unpublished Hong Kong Government data.

the effective legal boundaries of the People's Republic of China a business presence which will enable it to retain access to the international financial services on which it is dependent.

The second principal area of incompatibility is marketing. Most Hong Kong firms relocating to the People's Republic of China face the restriction that output will have to be exported to world markets ('Resolution of the Standing Committee of the NPC Approving Regulations on Special Economic Zones in Guangdong', 26 August 1960, Articles 9 and 17; 'Law of the People's Republic of China on Foreign Capital Enterprises', 12 April 1986, Article 3). The incompatibility here is that China, with its power and transport shortages, poor telecommunications and restrictions on foreign travel, news and publications, is not an ideal venue from which to engage in world trade (Chai 1983: 116–20).[25] Whereas 'Hong Kong's domestic advantages continue to be its unparalleled access to foreign markets as a regional trading centre. This implies Hong Kong support a wide range of service activities, including [those] which complement regional and world-wide trends' (CL Alexanders Laing and Cruickshank Securities (HK) Ltd 1988: 13). The result of this marketing incompatibility is similar to that attributed to financial incompatibility. Hong Kong firms must continue to maintain their links with world markets. The key management decisions about what to produce in response to export-market trends, together with the design, marketing, shipping and distribution facilities to serve overseas customers, must be retained in Hong Kong.

The physical production of goods, however, is an area of high compatibility. Hong Kong has relied heavily on the People's Republic of China over the years for a considerable proportion of retained industrial inputs and capital goods. The territory, until 1980, relied on the People's Republic of China for its additional labour (Youngson 1983; Sung 1990). The gap between the situations in Hong Kong and the People's Republic of China is smallest when it comes to physical production.

Chen's studies (1981 and 1983a) of outward investment in Hong Kong in the late 1970s and early 1980s suggest that Hong Kong investment in China differs from Hong Kong investment in other south-east Asian countries in that only the labour-intensive stages of production were transferred to China, and not the entire manufacturing process. This chapter gives the rationale for Hong Kong firms' decision to transfer manufacturing capacity to the People's Republic of China but to retain financial, marketing and central management operations in Hong Kong (Greenwood 1990: 272; CL Alexanders Laing and

Cruickshank Securities (HK) Ltd 1988). China has created incentives which will provide Hong Kong manufacturers (rather than any other group) with an environment that closely resembles that of Hong Kong itself. On the whole, however, the incentive package does not seem to have been a carefully engineered strategy designed to match Hong Kong's advantages as a business location item by item. It may be that the socialist-minded authorities lacked either the knowledge or the political willingness to adopt measures to replicate Hong Kong on Chinese soil. More likely, however, is that after removing the biggest obstacles to relocation to the People's Republic of China through the provision of a legal framework to protect the interests of Hong Kong firms and foreign investors,[26] the Central People's Government felt that enough had been done to 'pull' investors into the country. This assumption would have been all the more logical given the strength of the 'push' factor created by Hong Kong's labour shortage.

ECONOMIC INVOLVEMENT ACROSS THE BOUNDARY

New economic linkages have arisen between Hong Kong and China. Hong Kong now exports its value added to China, supplies capital for Chinese economic development and brings into China management, experience and technological knowledge (Sung 1990). A substantial amount of Hong Kong's manufacturing capacity has shifted across the border, although comparatively few statistical data are available on the exact extent of the transfer. On the Hong Kong side, the most recent (May 1991) comprehensive survey of manufacturers revealed that 28 per cent of Hong Kong manufacturing establishments had some form of outward-processing arrangement in Guangdong (Industry Department and Census and Statistics Department 1990: 34–5). Some 40 per cent of firms which already own plants in Guangdong had plans for further expansion. An additional 10 per cent of manufacturers who do not yet own plants in Guangdong had plans to expand operations there (Industry Department and Census and Statistics Department 1990: 34–6, 42–4). The findings of this survey are in keeping with Sit's 1987 study of Hong Kong's 'small and medium-sized'[27] manufacturing firms, in which 18 per cent of companies surveyed had 'out-processing facilities' (any form of investment or subcontracting arrangement) in Guangdong. Nearly half these firms reported that their Chinese operations generated over half their total output. Out of a sample of twenty-four larger firms which subcontracted within Hong Kong, 70

152

per cent reported also having out-processing arrangements with China (Sit 1989).

POLITICAL GUARANTEES FOR 1997

This chapter has suggested that China was able to attract Hong Kong manufacturers to relocate their productive capacity across the border due, first, to economic incentives and the relaxation of controls (the 'pull' factors) and, second, to the closing of the Hong Kong–China border to illegal immigration in 1979 (a 'push' factor). It is also suggested that manufacturing, amongst the various sectors of the Hong Kong economy, had the highest 'compatibility' with the political and economic systems of the People's Republic of China.

These general considerations were reflected in two major political acts. The first was the *Joint Declaration* (1984) and the second the *Basic Law* designed to give concrete expression to the *Joint Declaration*'s provisions. The *Joint Declaration* spelt out in considerable detail the guarantees made by the People's Republic of China in favour of Hong Kong continuing to be an integrated component of world markets. Hong Kong's boundaries will remain open to the rest of the world, so that unhindered communications and trade access to the world at large can be continued. Its financial and fiscal arrangements would remain undisturbed.[28] No attempt was made in the *Joint Declaration* to create opportunities to increase the areas of compatibility between Hong Kong and the People's Republic of China or to reduce the areas of incompatibility defined above.

The transfer of manufacturing capacity from Hong Kong to China in the 1980s without any shift in the associated financial and marketing facilities from Hong Kong reflected a state of affairs envisaged, at least implicitly, by the *Joint Declaration*. The behaviour of local business firms was in line with the state of affairs which the Central People's Government had anticipated in negotiations of the *Joint Declaration*: Hong Kong would continue to be a business centre with substantial insulation from those of China's systems and regulations that capitalism would find onerous, and yet would function as part of China's national system through its dependence on China both for imports and for its manufacturing capacity.

This selective opening of certain aspects of the Hong Kong–China boundary was the outcome of two factors. First, the Central People's Government tried to create a hospitable environment for outside investors, particularly through seeking to reassure the Hong

153

Kong business community.[29] Second, the manufacturing sector was encouraged by the labour shortage caused by the cessation of labour imports to transfer to the People's Republic of China.

The business generated by this 'portfolio' depends very heavily on the behaviour of the Central People's Government – for example, on the allocation of electricity and other scarce resources. Economically, the boundary between Hong Kong and China has begun to operate more like an internal Chinese boundary than the divisive boundary of 1949– 79. The Hong Kong economy has therefore become less insulated from the impact of political and economic changes within China. Even if the Central People's Government sought to insulate Hong Kong as a territory from domestic (Chinese) events, it would find this difficult unless similar insulation is offered to extensive areas of Guangdong Province. The barrier effects of the boundary which protected Hong Kong from much of the effects of Chinese policy have been eroded since 1978 and Hong Kong can no longer remain separate from the mainland. If the economic trends described in this chapter continue, Hong Kong's economy will become more closely tied to that of the mainland (Greenwood 1990; Peat Marwick Management Consultants Ltd 1989).

CONCLUSION

Miners 1980 was optimistic about Hong Kong's continued survival as a political entity separate from China. He based his view on the economic advantages China derived from the British colony. His discussion is

> based on the premise that China's policy makers will decide their future actions over Hong Kong on the basis of pragmatic calculations of the costs and benefits of exercising their legal right ... to take back the New Territories in 1997.
>
> (Miners 1980: 26)

This chapter has suggested that the policies China pursues towards its boundary with Hong Kong are revealing of the relationship it envisages after 1997. This chapter has showed that changed internal conditions in China and the Chinese strategy pursued towards Hong Kong have resulted in a new economic relationship between the two. The re-location of Hong Kong manufacturing industry across the border and the service functions performed for China by Hong Kong mean that Hong Kong can conceivably play an important role as part of the People's Republic of China. Chinese policies have prepared a way for the reabsorption of Hong Kong into the mainland by the 'one country, two systems' formula.

NOTES

1 Only Hong Kong Island (Treaty of Nanking, 1841) and Kowloon Peninsula (Convention of Peking, 1860) were ceded to Britain in perpetuity; these are now the main urban areas of Hong Kong. The New Territories (forming the major portion of Hong Kong's land area) were ceded under a ninety-nine-year lease (Convention of Peking, 1898) which expires in 1997. The rest of the territory is not considered viable without the New Territories, which is Hong Kong's main source of industrial land and local water supplies. Approximately three million people will be living in the New Territories by 1997.

2 The Sino-British 'Opium War' (1840–2); the 'Arrow War' (1858–60) and the Sino-Japanese War (1894–5).

3 'In contrast with the vacillating economic strategies and policies that prevailed in mainland China between 1952 and 1978, which often reflected ideological struggles over the proper objectives and means to implement socialism, Hong Kong maintained an extraordinary degree of stability in its political and economic institutions' (Rabushka 1990: 355).

4 Around 40 per cent of Hong Kong's population was born in China (*Hong Kong Yearbook* 1988: 302).

5 Article 1 of the *Constitution of the People's Republic of China* (1982): 'The People's Republic of China is a socialist state under the people's democratic dictatorship led by the working class and based on the alliance of workers and peasants. The socialist system is the basic system of the People's Republic of China. Disruption of the socialist system by any organization or individual is prohibited'.

6 China's desire to regulate contacts with Hong Kong, even during its 'Open Door' policy, is shown by its efforts to pull down Guangdong television aerials capable of receiving Hong Kong transmissions in order to curb the influence of the Hong Kong media (*Guangdong Ribao*, 9 April 1982; *Nanfang Ribao*, 14 June 1982; *Guangdong Radio Service*, 26 March and 4 June 1982, 6 January 1983).

7 Before 23 October 1980 immigrants entering Hong Kong illegally from China were allowed to stay if they had 'reached base' (that is, arrived in the urban areas of Hong Kong). Since then, illegal immigrants have been repatriated regardless of where they were found in the territory.

8 See Wong (1988) on the important contribution made to Hong Kong's economic development (in particular, the establishment of a modern textile industry and of shipping interests) by Shanghainese immigrants fleeing the communist regime in China.

9 These consist mainly of money, but also medicine and goods ranging from powdered milk to clothing. China has never published any official statistics on its non-trade earnings from Hong Kong; scholars and the Hong Kong government have only been able to estimate the magnitude of these earnings. Jao (1983: 36–42) has produced the most comprehensive estimates on the value of Hong Kong's remittances to China for 1950–80.

10 An English summary of the official Chinese version of these policy changes and of some of the relevant legislation can be found in Su (1985). See also Peng Zhen (1979) and (1982) and the Preamble of the *Constitution of the*

People's Republic of China (1982). English translations of the legislation cited in this chapter can be found in Legislative Affairs Commission (1987).

11 'Resolution of the Standing Committee of the NPC Approving the Regulations on Special Economic Zones in Guangdong Province' (16 August 1980) and 'Resolution of the Standing Committee of the NPC Authorizing the People's Congress of Guangdong and Fujian Provinces and their Standing Committee to Formulate Separate Economic Regulations for their Respective Special Economic Zones' (26 November 1981). An English translation of these laws can be found in Legislative Affairs Commission (1987).

12 See 'Regulations of the Criteria for Defining Urban and Rural Areas' (1956) and 'Regulations for Household Registration of the People's Republic of China' (1959), both cited in Potter and Potter (1990: 301–2); and Potter and Potter (1990) on the effects on a Guangdong village of household registration and of the restrictions on population movements, 300–6 (on pre-1970 controls) and 311–12 (on relative liberalization in 1979–85).

13 Huang Hua (1972), letter to UN Special Committee on Colonialism, 8 March 1972.

14 In 1978, for example, the Chinese authorities protested against the Hong Kong government's alleged attempt to deport Fu Chi and Shek Wai from Hong Kong.

15 These agreements are confidential; neither the texts nor information on the negotiation process are available, although official exchanges during the 'Yang Yang' incident (see note 17) give some indication of the nature of certain agreements. A letter from William Ehrman, Political Advisor to the Governor of Hong Kong, suggests that British authorities left adequate loopholes for exceptional cases – notably asylum arrangements.

16 See *Shun Pao*, 18 February 1987, for a report on Chinese officials' views. Chinese sensitivity on this issue was most clearly shown after the suppression of the pro-democracy movement in 1989. Since then, China has repeatedly warned Hong Kong against becoming a 'subversive base' and an 'anti-Communist base'.

17 The Yang Yang incident supports this view. In August 1989 a swimmer from China named Yang Yang requested political asylum in Hong Kong as he was afraid that China would punish him for having taken part in pro-democracy activities in Hong Kong. After considerable dispute, both within Hong Kong and with China, Yang was sent on to America where he had been offered asylum. China then claimed that Hong Kong had violated a 1982 agreement to repatriate 'two-way permit holders' (for a fixed term) who over-stayed in the territory, and that consequently China was no longer bound to accept the repatriation of illegal immigrants.

18 Before 1979 Chinese residents could come to Hong Kong by applying to the Chinese authorities for Chinese exit permits or to the Immigration Department, Hong Kong, for entry permits.

19 Classified Hong Kong Government document, 1982.

20 Numbers peaked between 15 January 1979 and 9 March 1979 (just before new immigration regulations came into effect), when 98,000 application

forms for entry visas for permanent residence in Hong Kong were issued. Some 20,000 completed forms were returned by 9 March 1979, of which 98 per cent met residence qualifications (confidential information note, Hong Kong Government, February 1990).

21 For legal definitions of Chinese nationals, see 'Nationality Law of the People's Republic of China', 10 September 1980, especially Article 4. On the political and legal dangers facing Hong Kong people in China and the less-attractive aspects of the Chinese legal system, see Fang 1986; Leung 1986; Li 1986; and Yu 1986. A discussion of the death sentence on a Hong Kong businessman convicted of smuggling in China and the possible adverse effects upon Hong Kong-China trade is found in Tsai 1986.

22 'The sheer spread and magnitude of the move across the border implies that manufacturers were not all looking to expand, but were being driven offshore seeking lower costs and more stable supplies of labour': Peat Marwick Management Consultants Ltd 1989: 5.

23 Increased illegal immigration from 1976 onwards reflected a trend towards greater liberality in China after the end of the Cultural Revolution and the fall of the Gang of Four.

24 'China's financial structure is incompatible, in many respects, with economic development. The main problem is that the mechanism of raising, distributing and managing capital funds is outdated' (Liu 1990: 125-6). See also Chai 1983: 119 and 151; 'Law of the People's Republic of China on Foreign Capital Enterprises', 12 April 1986, Articles 18 and 19, and May 1990, Chapters 3 and 4.

25 It was reported in 1990 that there were still power shortages for both local and foreign-invested enterprises in Guangdong Province, despite a sixfold increase in power-generating capacity between 1979 and 1989 (Hong Kong Trade and Development Council 1990: 5).

26 New laws enacted to cater for outside investment and trade include 'Law of the People's Republic of China on Chinese-Foreign Equity Joint Ventures', 8 July 1979; 'Economic Contract Law of the People's Republic of China', 1 July 1982; 'Trademark Law of the People's Republic of China', 3 March 1983; 'Law of the People's Republic of China on Economic Contracts Involving Foreign Interest', 1 July 1985; and 'Law of the People's Republic of China on Foreign Capital Enterprises', 12 April 1986.

27 These account for 94 per cent of the total of Hong Kong manufacturing firms.

28 See Section 3 (5-10); Annex I (I, V, VI, VII, IX) of the *Joint Declaration of the Government of the United Kingdom of Great Britain and Northern Ireland and the Government of the People's Republic of China on the Future of Hong Kong* (1984); and Articles 4, 6, 7, 8, 10, 11, 15, 58, 104-41 of the *Basic Law of the Hong Kong Special Administrative Region of the People's Republic of China.*

29 Two main approaches were used: first, through the official political and legal provisions of the *Joint Declaration* and the *Basic Law*; and second, through a campaign to woo businessmen, conducted at both the public and personal levels (Lu 1982); *Zhongguo Xinwen She* 0830 GMT 25 November 1983; *South China Morning Post*, 30 October 1982, and *Hong*

Kong Standard, 25 November 1983, cited in Scott (1989: 174). Scott (1989: 198) notes that top officials of the New China News Agency went out of their way to entertain and reassure investors. The new director of the New China News Agency was reported to dine 'nearly every night with bankers and businessmen ... relaying Peking's message that there is no need to worry, that there will be only minimal changes after China recovers sovereignty and that China really wants China's capitalist system to continue' (Lee 1983).

REFERENCES

Newspapers and periodicals

Far East and Economic Review
Guangdong Radio Service
Guangdong Ribao
Hong Kong Times
Journal of Oriental Studies
Ming Pao
Nanfang Ribao
Shun Pao
South China Morning Post
The Hong Kong Standard
The Nineties
The Seventies
Wen Wei Pao
Zhongguo Xinwen She

Government reports and statutes

Almanac of China's Economy (various years) Beijing: Economic Management Press.
Almanac of China's Foreign Economic Relations and Trade (annual publication) Hong Kong: China Resources Trade Consultancy Co. Ltd.
Basic Law of the Hong Kong Special Administrative Region of the People's Republic of China, Basic Law Consultative Committee, Hong Kong.
Census and Statistics Department (1990) *Hong Kong Social and Economic Trends 1978–1988*, Hong Kong: The Government Printer.
China's Foreign Economic Legislation (1982) vol. I, Beijing: Foreign Languages Press.
China's Foreign Economic Legislation (1986) vol. II, Beijing: Foreign Languages Press.
China Statistical Yearbook (various years) (in Chinese) Beijing.
China Statistical Yearbook 1989 (1990) Beijing: China Statistical Information and Consultancy Service Centre and International Centre for the Advancement of Science and Technology Ltd.
Constitution of the People's Republic of China (1982), Beijing: Foreign Languages Press.

Hong Kong Government (annual publication) *Hong Kong Annual Reports*, Hong Kong: The Government Printer.

Hong Kong Hansard (annual publication) Hong Kong: The Government Printer.

Immigration Department (annual publication) *Annual Reports*, Hong Kong: The Government Printer.

Industry Department and Census and Statistics Department, Hong Kong Government (1991) *Survey on the Future Development of Industry in Hong Kong, Statistical Survey of Manufacturers*, Hong Kong: The Government Printer.

Joint Declaration of the Government of the United Kingdom of Great Britain and Northern Ireland and the Government of the People's Republic of China on the Future of Hong Kong (1984) Hong Kong: Government Printer.

Legislative Affairs Commission of the Standing Committee of the National People's Congress (1987) *The Laws of the People's Republic of China 1979–82*, Beijing: Foreign Languages Press.

Legislative Affairs Commission of the Standing Committee of the National People's Congress (1987) *The Laws of the People's Republic of China 1983–86*, Beijing: Foreign Languages Press.

Peng Zhen (1979) 'Explanations of the Seven Draft Letters', delivered at the 2nd Session of the 5th National People's Congress, 26 June 1979, in Legislative Affairs Commission of the Standing Committee of the National People's Congress (1987) *Laws of the People's Republic of China, 1979–82*, Beijing: Foreign Languages Press.

Peng Zhen (1982) 'Report on the Draft of the Revised Constitution of the People's Republic of China', delivered at the 5th Session of the 5th National People's Congress, 26 November 1982, in Legislative Affairs Commission of the Standing Committee of the National People's Congress (1987) *Laws of the People's Republic of China, 1979–82*, Beijing: Foreign Languages Press.

Royal Hong Kong Police Force (annual publication) *Annual Review*, Hong Kong: The Government Printer.

Secondary sources

Chai, J. (1983) 'Industrial Co-Operation between China and Hong Kong', in A.J. Youngson (ed.), *China and Hong Kong: The Economic Nexus*, Oxford: Oxford University Press, 105–55.

Chau, L.C. (1983) 'Imports of Consumer Goods from China and the Economic Growth of Hong Kong', in A.J. Youngson (ed.) *China and Hong Kong: The Economic Nexus*, Oxford: Oxford University Press, 184–225.

Chen, Edward, K.Y. (1981) 'Hong Kong Multinationals in Asia: Characteristics and Objectives', in K. Kumar and M.G. Macleod (eds), *Multinationals from Developing Countries*, Lexington, MA: Lexington Books.

—— (1983a) 'Multinationals from Hong Kong', in S. Lall (ed.), *The New Multinationals: The Spread of Third World Enterprises*, London: Wiley.

—— (1983b) 'The Impact of China's Four Modernizations in Hong Kong's Economic Development', in A.J. Youngson (ed.), *China and Hong Kong: The Economic Nexus*, Oxford: Oxford University Press, 77–103.

—— (1984) 'The Economic Setting', in D. Lethbridge (ed.), *Hong Kong: The Business Environment*, Oxford: Oxford University Press, 1–51.

CL Alexanders Laing and Cruickshank Securities (HK) Ltd (1988) 'Hong Kong's Economic Transformation: Changing Gear Again', Hong Kong.

Dicks, A. (1983) 'Treaty, Grant, Usage or Suffrance? Legal Aspects of the Status of Hong Kong', *China Quarterly*: 95.

Dorn, J.A. and Wang Xi (eds) (1990) *Economic Reform in China*, Chicago: The University of Chicago Press.

Fang Su (1986) 'Legal Protection for Hong Kong People on the Mainland' (in Chinese), *The Nineties*, 198: 55–6.

Far East and Economic Review (1969) *Asia Yearbook*, Hong Kong: FEER.

Feng Bangyan (1989) 'The Role of Hong Kong in the Course of China's Modernization' (in Chinese), *Jingji Yanjiu*, 4: 64–70.

Fu Huichang (1989) 'Open up the Mountain Gate, Get into the City Gate, and Stride out of the National Gate – Exploring Ways for Promoting Guangdong's Economic Development' (in Mandarin), talk delivered 10 May on International News and Current Events programme, Beijing Domestic Service.

Greenwood, J.G. (1990) 'The Integration of Hong Kong and China', in J.A. Dorn and Wang Xi (eds), *Economic Reform in China*, Chicago: The University of Chicago Press, 125–7; 271–6.

Heng Seng Bank (1987) *Heng Seng Economic Monthly*, April.

—— (1990) *Heng Seng Economic Monthly*, March.

Hoadley, J.S. (1970) 'Hong Kong is the Lifeboat: Notes on Political Culture and Socialisation', *Journal of Oriental Studies*, 8: 206–18.

Hong Kong Trade and Development Council Research Department (1990) *Recent Investment Environments of Guangdong, Fujian and Hainan*, Hong Kong: Hong Kong Trade and Development Council.

Hsu, I. (1982) *The Rise of Modern China*, Hong Kong: Oxford University Press.

Hsueh Tien-tung and Woo Tun-oy (1981) 'Trade between Hong Kong and China: Issues and Prospects', Working Paper no. CC12, Centre of Asian Studies, University of Hong Kong: Hong Kong.

Hung, C.L. (1984) 'Foreign Investments', in D. Lethbridge (ed.), *Hong Kong: The Business Environment*, Oxford: Oxford University Press, 180–210

Jao, Y.C. (1980) 'Hong Kong as a Regional Financial Centre: Evolution and Prospects', in C.K. Leung (eds), *Hong Kong: Dilemmas of Growth*, Botany: Australian National University Press: 161–94.

—— (1983) 'Hong Kong's Role in Financing China's Modernization', in A.J. Youngson (ed.), *China and Hong Kong: The Economic Nexus*, Oxford: Oxford University Press, 12–76.

—— (1984) 'The Financial Structure', in D. Lethbridge (ed.), *Hong Kong: The Business Environment*, Oxford: Oxford University Press, 124–70

Kumar, K. and Macleod, M.G. (eds) (1981) *Multinationals from Developing Countries*, Lexington, MA: Lexington Books.

Lall, S. (1983) *The New Multinationals: The Spread of Third World Enterprises*, London: Wiley.

Lane, K.P. (1990) *Sovereignty and the Status Quo. The Historical Roots of China's Hong Kong Policy*, Boulder: Westview Press.

Lau Siu-kai (1982) *Society and Politics in Hong Kong*, Hong Kong: The Chinese University Press.

Lau Siu-kai and Kaun Hsin-chi (1988) *The Ethos of the Hong Kong Chinese*, Hong Kong: The Chinese University Press.

Lee, M. (1983) 'Hearts and Minds – and a Barrage of Words', *Far Eastern and Economic Review*, 25 August.

Lethbridge, D. (ed.) (1984) *Hong Kong: The Business Environment*, Oxford: Oxford University Press.

Lethbridge, H. (1984) 'The Social Structure: Some Observations', in D. Lethbridge (ed.), *Hong Kong: The Business Environment*, Oxford: Oxford University Press, 52–69.

Leung Chi-Keung, Cushman, J.W. and Wang Gungwu (eds) (1980) *Hong Kong: Dilemmas of Growth*, Botany: Australian National University Press.

Leung, Woren (1986) 'Because of a Contract Dispute, He was Seized and Held to Ransom – Hong Kong Businessman Reveals his Experience as a Shanghai Prisoner' (in Chinese), *The Nineties*, 198: 48–51.

Li Yi (1986) 'Hong Kong People Held in Custody; To whom can they Turn for Help?' (in Chinese), *The Nineties*, 198: 42–4.

Liu Funian (1990) 'Financial Reform, a Prerequisite for Development', in J.A. Dorn and Wang Xi (eds), *Economic Reform in China*, Chicago: The University of Chicago Press, 125–7.

Lu Keng (1982) 'Hong Kong Self-Rule by the Hong Kong People' (in Chinese), *Baixing*, 34.

Ma Hong (ed.) (1990) *Modern China's Economy and Management*, Beijing: Foreign Languages Press.

Miners, N. (1975) 'Hong Kong: A Case Study in Political Stability', *Journal of Commonwealth and Comparative Politics*, 13: 26–39.

—— (1986) *The Government and Politics of Hong Kong*, Hong Kong: Oxford University Press.

Moser, Leo J. (1985) *The Chinese Mosaic: The Peoples and Provinces of China*, Boulder: Westview Press.

Peat Marwick Management Consultants Ltd (1989) *The Restructuring of Hong Kong's Manufacturing Sector. A Critical Transformation*.

Potter, S.H. and Potter, J. (1990) *Chinese Peasants, The Anthropology of a Revolution*, Cambridge: Cambridge University Press.

Rabushka, A. (1979) *Hong Kong: A Study in Economic Freedom*, Chicago: University of Chicago Press.

—— (1990) 'A Free Market Constitution for Hong Kong: A Blueprint for China', in J.A. Dorn and Wang Xi (eds), *Economic Reform in China*, Chicago: The University of Chicago Press, 335–45.

Schiffer, J.R. (1983) *Anatomy of a Laissez-Faire Government: The Hong Kong Growth Model Reconsidered*, Hong Kong: Centre of Urban Studies and Urban Planning, University of Hong Kong.

Scott, I. (1989) *Political Change and the Crisis of Legitimacy in Hong Kong*, London: Hurst and Company.

Sit, V.F.S. (1989) *Hong Kong's Industrial Out-Processing in the Pearl River Delta of China*, Hong Kong: Centre of Asian Studies, University of Hong Kong.

161

Su Wenming (ed.) (1985) 'The Open Policy at Work', *China Today* (15), Beijing: Beijing Review Publication.

Sung Yun-wing (1990) 'The China–Hong Kong Connection', in J.A. Dorn and Wang Xi (eds), *Economic Reform in China*, Chicago: The University of Chicago Press, 255–66.

Tsai Xin (1986) 'An Examination of China–Hong Kong Trade in the Context of the Zhang Zhenwei Case' (in Chinese), *The Nineties*, 198: 52–4.

Wesley-Smith, P. (1980) *Unequal Treaty 1898–1997*, Hong Kong: Oxford University Press.

Wong Siu-lun (1988) *Emigrant Entrepreneurs: Shanghai Industrialists in Hong Kong*, Hong Kong: Oxford University Press.

Youngson, A.J. (ed.) (1983) *China and Hong Kong: The Economic Nexus*, Oxford: Oxford University Press.

Yu Zawen (1986) 'On the Mainland, How Many Hong Kong People Have Been Sentenced?' (in Chinese), *The Nineties*, 198: 45–7.

9

HOPPO RYODO MONDAI
(The Northern Territories Problem)
A territorial issue between Japan and Russia

Kimie Hara

INTRODUCTION

Japan has three unresolved boundary problems that affect its territorial sovereignty. They are the Senkaku Retto, Takeshima and Hoppo Ryodo problems, respectively with China, South Korea and the former Soviet Union – currently Russia, a member of the Commonwealth of Independent States (CIS).

With respect to Senkaku Retto (*Tiao-yu* in Chinese)[1] – a small chain of islands claimed simultaneously by Japan, China and Taiwan – Japan and China agreed to freeze the status quo and entrust resolution of the issue to the next generation. This policy stance is based on the broad perspective that the territorial issue should not be an impediment to the development of friendly relations between the two countries. Through the adoption of this policy, the territorial question was put aside when normal relations were established in 1972, and when the Japan–China Treaty of Peace and Friendship was signed in 1978. Japan does not recognize Taiwan as a negotiating partner for this issue.

Japan disputes the ownership of Takeshima (*Tok-do* in Korean)[2] with South Korea, and the two nations did not reach any agreement when their relations were normalized in 1965. At the tenth annual cabinet meeting between Japan and South Korea in 1978, it was jointly decided to continue negotiations regarding ownership, but the dispute has not been settled yet. Since 'no third party interest or strategic security concern exists to explain the sensitivity of this issue' (Johnston and Valencia 1990: 113–14), except for a small petroleum potential, it has a low priority on the diplomatic agenda.

The 'Northern Territories problem' (*Hoppo Ryodo Mondai*) is the

163

issue Japan has most adamantly pursued regarding its sovereignty. Japan has resolutely insisted on the return of the northern islands as 'inalienable' territories, which have been occupied by the Soviets since the end of World War II. Today this territorial problem is still unresolved, although it is no longer the Soviet Union, but Russia, that is a party to negotiation. The issue remains to this day the biggest obstacle preventing the two nations from signing a peace treaty and improving relations. The Northern Territories issue is the product of a complicated mixture of factors involving history, politics, strategic concerns and economics. Considering the past and present involvement of the United States in the issue, it may not necessarily be resolved through strictly bilateral negotiations. With the accelerating pace of change in the current world situation, especially within the former Soviet Union, political circumstances affecting the solution of the issue are also changing. This chapter focuses on the 'Northern Territories', and the nature of the issue, its development and prospects for the future will be discussed from several perspectives. The rival claims of Japan and Russia are summarized in Table 9.1.

BACKGROUND OF THE DISPUTED ISLANDS

South-west of Russian Kamchatka and the Sea of Okhotsk, the 'Northern Territories', as they are collectively referred to by the Japanese, are comprised of the four islands, or island groups, of Etorofu, Kunashiri, Shikotan and the Habomai islands. Beginning only 3.7 kilometres off the tip of Cape Nosappu of Hokkaido, the islands stretch in a north-easterly direction, comprising 4,996 square kilometres of land, with Kunashiri and Etorofu accounting for about 90 per cent of the total land area. By the Russians the disputed islands are considered to be the southernmost islands of the Kuriles, a Russian-held archipelago that stretches for a thousand kilometres between Hokkaido and the Kamchatka Peninsula. The Japanese government, however, claims that these islands are distinct from the Kuriles.

Despite their northern location, these islands have a relatively mild climate. The waters around the islands, at the confluence of warm and cold currents, abound in fish and other marine products, and make up what is considered to be one of the three best fishing-grounds in the world. The main industry on these islands, where Russians are living today, is fishing, as it was during Japanese rule.

As a part of the gateway of islands between the Pacific Ocean and the Sea of Okhotsk, the disputed territories have long been strategically

Table 9.1 Rival claims to the Northern Territories

THE JAPANESE CLAIMS

(1) The Treaties of Shimoda (1855) and St Petersburg (1875) confirmed that Kunashiri, Etorofu, Shikotan and the Habomai islands are inalienable Japanese territories.

(2) The Yalta Agreement (1945), which gave the islands to the Soviets, was an illegitimate secret agreement among the Allies (USSR, USA and UK) to which Japan was not a party. Japan did not know the Agreement even existed. The Cairo Declaration (1943) and Potsdam Declaration (1945) stated that the Allies (including the USSR) were not fighting for territorial gain, so the Soviet seizure of the Northern Territories was illegitimate.

(3) The Northern Territories are not part of the Kuriles, which Japan gave up in the San Francisco Peace Treaty (1951). Moscow is, anyway, not entitled to pursue claims under the San Francisco Treaty since it failed to sign it.

(4) The Soviets agreed to the return of Shikotan and the Habomai islands in the Japan–Soviet Joint Declaration (1956).

THE RUSSIAN CLAIMS

(1) The treaties of 1855 and 1875 lost their validity as a consequence of the Russo-Japanese War (1904) and the Portsmouth Treaty (1905).

(2) The Yalta Agreement and Potsdam Declaration are legitimate and represent the final solution to the territorial issues.

(3) Japan started the war – and lost. It must therefore accept the consequences of defeat.

(4) In signing the San Francisco Peace Treaty, Japan gave up its claim to the Kuriles.

(5) The Kuriles do include the disputed islands.

---

important to both nations. Etorofu, for example, served as the staging area for the task-force of the Japanese Imperial Navy at the time of the Pearl Harbor attack. For the Russians, these islands are located so as to allow control of the sea-lanes between the Sea of Okhotsk and the Pacific Ocean. They are located on a major pathway for Russian submarines operating out of the Sea of Okhotsk, where much of their sea-based nuclear force is concentrated. According to Defence of Japan, '91, ground forces equivalent to a division have been deployed on Etorofu, Kunashiri and Shikotan islands since 1978. When Soviet President Gorbachev visited Japan in April 1991, he made a proposal to reduce the forces stationed on these islands, as stated in the Joint Communiqué. However, details of the reduction are still unknown (Ministry of Defence, Japan, 1991: 53–4).

Geographically, Russia is Japan's closest neighbour. Geographical proximity in Asia has generated many confrontations over boundaries, rather than creating friendly relationships in the past: Japanese–Soviet/ Russian relations are not an exception to this. Russia has historically appeared as a northern 'threat' to Japan, rather than a friendly neighbour. Beginning in the eighteenth century, there has been a long history of military hostility between the two countries. Their rivalry to extend influence in north-east Asia led to actual military clashes on several different occasions in the twentieth century. However, the historical dispute over the sovereignty of the 'Northern Territories' goes back to an even earlier time. Both countries claim first 'discovery' of these islands. It was during the seventeenth to eighteenth centuries that Japan and Russia began to reach the islands stretching between Hokkaido and Kamchatka. Long before their discovery by Japan or Russia, however, these islands were populated by native people called the Ainu. As the names of these islands are derived from the language of the Ainu, there is no doubt that they originally belonged to neither Japan nor Russia, but to the Ainu.

In 1855, a year after Japan ended her history of over 200 years of national isolation by signing the Treaty of Kanagawa with the United States, Imperial Russia and Tokugawa Japan signed the first treaty in history between the two countries – the Treaty of Commerce, Navigation and Delimitation (Treaty of Shimoda). This set a boundary between Etorofu and Uruppu, and stipulated that the Kurile islands north of Uruppu belonged to Russia. The treaty also stipulated that the large island of Sakhalin north of Hokkaido would have no national boundary but would remain open to settlement by both nations.

In 1875 with the Treaty for Exchange of Sakhalin for the Kurile Islands (Treaty of St Petersburg) the Meiji government abandoned all of Sakhalin in exchange for the rights to the Kurile chain. Article 2 of this treaty lists the names of eighteen islands of the Kuriles, from Shimushu to Uruppu, which were to be handed over to Japan (Ministry of Foreign Affairs, Japan, 1987: 5). This was the last time the two countries defined their respective territories as the result of a peaceful negotiation. In 1904 Japan defeated Russia in the Russo-Japanese War. As a result, Southern Sakhalin was given to Japan by the Portsmouth Treaty of 1905. Since then, relations between the two countries continued to deteriorate, and there followed several military clashes: the Japanese intervention in Siberia 1918–22, the Changkufang incident of 1938, the Nomonhan incident of 1939 and the last stage of World War II.

On 15 August 1945 Japan accepted the Potsdam Declaration and surrendered to the Allies. Just six days earlier, on 9 August, the Soviet Union entered into a state of war against Japan in violation of the

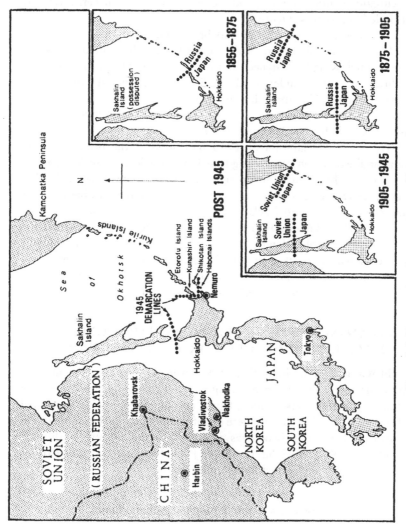

Figure 9.1 Demarcations between Japan and Russia

Neutrality Pact, which was signed with Japan in 1941. On 18 August, three days after Japan's surrender, the Soviet army landed in the northernmost of the Kuriles (then Japanese territory) and began attacking Japanese positions. By 3 September Soviet troops had occupied all the islands between Kamchatka and Hokkaido. The occupation continues today, and the disputed 'Northern Territories' are included in this area (Figure 9.1).

The Soviet move on Japan had been arranged earlier, in February 1945 at Yalta. The leaders of the Soviet Union, the United States and the United Kingdom signed a secret protocol which outlined the political conditions under which the Soviet Union would enter the war against Japan. The three leaders agreed that the southern part of Sakhalin should be 'returned' to the Soviet Union and that the Kurile Islands should be 'handed over' to the Soviet Union.

The Cairo Declaration of 1943 had outlined the principle of no territorial expansion, stipulating that Japan would be expelled from all territories it had taken 'by violence and greed'. The Potsdam Declaration, which Japan accepted at the time of surrender, was basically the successor to the Cairo Declaration. Later, in the San Francisco Peace Treaty of 1951, Japan was further forced to renounce 'all right, title and claim to the Kurile Islands, and to that portion of Sakhalin and the islands adjacent to it over which Japan acquired sovereignty as a consequence of the Treaty of Portsmouth'. As a consequence, over 17,000 Japanese were expelled from the disputed islands.

JAPAN–SOVIET RELATIONS IN THE COLD WAR ERA: TWO CHANCES FOR *RAPPROCHEMENT*

Cold War geopolitics in north-east Asia were in many respects merely a continuation of the historical triangular competition between the three major regional powers – Japan, China and the USSR – with the United States now added to the Japanese leg of the triangle. In this context, the Northern Territories issue was generated as an extremely difficult problem to solve, which has prevented Japan from signing a peace treaty and improving relations with the Soviet Union.

Hatoyama's visit to Moscow: the first chance

Following World War II, when the world was divided into East and West with the Soviet Union and the United States as the superpowers in a bipolar system, Japan found it most beneficial to put itself under the

American umbrella through the security arrangement. During the first decade after the war (1945–55), there were no diplomatic relations between Japan and the Soviet Union.

The first chance for *rapprochement* came in the mid-1950s, which was highlighted with Prime Minister Hatoyama's visit to Moscow. The death of Stalin had created potential for political change in the USSR. Soviet political leaders – as became apparent at the 20th Party Congress – were seeking new directions for Soviet diplomacy and promoting 'peaceful co-existence' abroad. Hatoyama, who sought Soviet agreement for Japan to join the United Nations and the return of Japanese POWs who were still detained in Siberia, accepted the Soviet overtures (Hatoyama 1957: 178). As a result, diplomatic relations between Moscow and Tokyo were restored, and in the Joint Declaration which followed the visit both countries agreed to transfer Shikotan and the Habomai islands to Japan.

It is about this time that the territorial issue became more complicated, including domestic political machinations. Japan and the USSR were on the verge of reconciliation with the transfer of the two small island groups to Japan, but the Japanese government was deeply divided. Japanese conservatives led by former Prime Minister Yoshida placed priority on diplomatic co-operation with the United States, which desired that Japan avoid any reconciliation with the Soviet Union. Under pressure from these elements, the Hatoyama cabinet added Etorofu and Kunashiri to Japanese claims in the course of the negotiations. Since then the Japanese government has steadfastly insisted on the return of all four island groups.

Later, in 1960, during a period of deteriorating East–West relations, the Kishi cabinet agreed to a revised US–Japan Security Treaty. The Soviet response was Gromyko's announcement that Shikotan and the Habomai islands would not be transferred until all foreign troops had completely withdrawn from Japan. Since then, relations chilled deeply.

Tanaka's visit to Moscow: the second chance

A second chance to normalize Soviet–Japanese relations came about in the early 1970s with the rising confrontation between China and the USSR, and the growth of *détente* between the United States and USSR. Japan had achieved rapid economic growth despite her defeat in the war, and the Soviets were beginning to see Japan as an attractive capitalist country for economic exchange.

Prime Minister Tanaka visited Moscow in 1973, but in the Joint

Communiqué that followed there was no clear statement regarding territorial problems. It merely said that 'the two sides recognized that to conclude a peace treaty by resolving the yet unresolved problems remaining since World War II would contribute to the establishment of truly good-neighbourly relations'. Although Japanese officials say Mr Brezhnev twice concurred with Mr Tanaka's confirmation that 'the unresolved problems' included the 'Northern Territories problem' during their summit conference (Ministry of Foreign Affairs, Japan, 1987: 14), it was later denied by the Soviet side. Even in this period of East–West *détente*, there was no clear progress in Japanese–Soviet relations.

The signing of the Japan–China Treaty of Peace and Friendship in 1978, incorporating an 'anti-hegemony' clause directed against the USSR at Chinese insistence, had an immediate effect on Soviet–Japanese relations. The Soviets began a build-up of forces in the Far East, including the deployment of 40 MiG-23 fighter-planes and one division of troops and the construction of new military facilities on Etorofu and Kunashiri. Relations deteriorated further in 1979 when Japan joined the United States in sanctions against the Soviet Union over the latter's invasion of Afghanistan, and regular foreign ministerial contact was suspended between the two nations.

At the time of the Soviet invasion to Afghanistan, Japan for the first time specified the USSR as a potential threat in a White Paper prepared by the Japanese defence agency. Furthermore, Soviet deployment of the Yanky-III-class SLBM submarines, which could attack the American mainland from the Sea of Japan, increased the strategic importance of the region in the late 1970s. Since a key element of Soviet strategy was to defend their Pacific bases and maintain access to the Pacific, the USSR rapidly increased its Pacific fleet, air-force and navy, in addition to their deployment in the 'Northern Territories'.

Meanwhile, the Japanese government designated a 'Northern Territories Day'[3] in 1981, and promoted a campaign to make the return of the Northern Territories 'the unanimous wish of the entire nation'. Since the mid-1970s the Soviet response to the question of the territorial issue was consistently cool, and the official position was that 'no territorial problem exists between Japan and the Soviet Union'.

Involvement of the United States

As was suggested earlier, when considering the Northern Territories issue in connection with global international politics, one cannot neglect the role of the United States. The United States has made this issue more complicated and difficult to solve. A careful review of US involvement in the territorial issue is important in defining the nature of the problem and also in finding a solution. If one looks back over history, US attitudes toward these islands have not been consistent. The United States has changed its policy on the treatment of these islands on several occasions, depending on its national interest in the context of international politics.

In US wartime policy, these islands were little more than bargaining chips to help bring the Soviets into the war in the Far East. At the time of the Yalta Agreement, by which transfer of the Kuriles to the USSR was promised, the military situation required the United States to ask Soviet co-operation in fighting the Japanese. However, as the military situation became advantageous to the United States toward the end of the war, it no longer seemed interested in passing these islands into Soviet hands. When Truman sent a message to Moscow on 15 August 1945, explaining the terms of Japanese capitulation, the Kuriles were excluded from the area of Soviet occupation (the dispossession of Manchuria, of northern Korea and of Sakhalin was mentioned); but due to Stalin's strong objection, the United States agreed to include the Kuriles in the area of Soviet occupation (Irie 1962: 134).

After the Japanese surrender, the United States had gained a new sphere of influence in the Pacific – Micronesia – by tactically negotiating their position on the Kuriles. When conflicts of interest arose between the United States and the USSR over the post-war treatment of Micronesia, US Secretary of State Byrnes used the Kuriles for political leverage. Byrnes told his Soviet political counterpart, Molotov, 'that Mr Roosevelt has said repeatedly at Yalta that territory could be ceded only at the peace conference and he had agreed only to support the Soviet Union's claim at the conference' (Byrnes 1947: 221). He implied the ultimate disposition of the Kuriles and the southern half of Sakhalin would be decided at a future peace conference, depending on the Soviet attitude toward the US trusteeship. Molotov grasped the implication, and when the US trusteeship agreement was voted upon later by the UN Security Council the USSR did not use its veto.

The United States forced Japan to renounce the Kurile islands, together with southern Sakhalin, by signing the San Francisco Peace Treaty. However, the Soviet Union did not sign it; and the treaty did not specify to whom these territories were renounced.

During the Cold War the primary American diplomatic objective regarding Japan was to prevent it from approaching the communist bloc. In order to maintain the US–Japanese security treaty and America's military presence in Japan, the United States did not want any *rapprochement* between Japan and the USSR, which would interfere with US global strategy. Therefore, when it appeared that Japan might make concessions on the territorial issue, it was seen as against American interests. The Japanese claim needed to be made unacceptable for the USSR, so that the territorial problem would remain an obstacle. In 1955 (prior to Hatoyama's visit to Moscow) the US Secretary of State, John Foster Dulles, put pressure on Japan by warning that Japan's potential sovereignty over Okinawa would be endangered if Japan were to make any concessions to the USSR.

After the Japanese territorial claim was revised, the Pentagon chose different tactics by declaring its support for this policy, and thus made the problem more complicated. In fact, as the Cold War escalated and the strategic importance of the Kuriles increased, it was reasonable for the United States to support the claim of its ally, Japan, to sovereignty over these islands; but it was convenient for the United States, too, to have the USSR keep the archipelago so that Japanese hostility toward the USSR could be maintained.

Range of the Kuriles

The Kuriles were not geographically defined in the San Francisco Treaty. But when the United States announced its legal position in 1956 in an *aide-mémoire* to Japan, it was stated that Kunashiri and Etorofu – together with the Habomai islands and Shikotan, considered a part of Hokkaido – have long been part of Japanese territory and that they should be justly acknowledged as under the sovereignty of Japan. When the United States reaffirmed this position in a note sent to the USSR in 1957 (concerning the shooting-down of a US aircraft over Hokkaido by the Soviets in 1954), it said the reference to the 'Kurile islands' in the San Francisco Peace Treaty and Yalta Agreement did not include, nor was intended to include, the Habomai islands, Shikotan, Kunashiri and Etorofu (Ministry of Foreign Affairs, Japan, 1987: 11). These statements were clearly the result of a politically inspired interpretation. Several State Department documents which verify this view were found in the National Archives in Washington at the end of 1990. These documents outline various plans for the future of the islands proposed in the Pentagon in the years leading up to the peace conference. The plans ranged from recognizing Soviet entitlement to all four islands, or giving the four islands to Japan, to giving Japan only two of them. These plans,

172

including the political justifications supplementing them, explain that US policy regarding the peace treaty with Japan was becoming more and more complex as US–Soviet relations worsened due to the Cold War (*Asashishinbun-sha* 1991: 147–71).

To this day there are disputes about the geographical definition of the Kuriles. The official Japanese position that these four islands are distinct from the Kuriles is based largely on these US pronouncements. As another major argument, the Japanese point out that the names of the eighteen islands making up the Kuriles were specified in the Treaty of St Petersburg (1875). However, a recent study by a Japanese scholar suggested this interpretation cannot be drawn from the original Dutch text (*Asashishinbun-sha* 1991: 155).

The Russians may have been more consistent as far as their interpretation of the Kuriles is concerned. The rest of the world has not paid much attention to the geographical classification of small islands like these in the Far East. One way to view their recognition, but avoid the political machinations of recent times, is to look at historical descriptions. For example, in the explanation of the 'Kurile islands' in the *Encyclopaedia Britannica* (1882), Iturup (Etorofu) and Kunashir (Kunashiri) are specified as some of the principal islands of the Kuriles, but Shikotan and the Habomai islands are not mentioned. However, as we have already seen, the problem is not easily solved by a clear definition of the Kuriles. The history of the Northern Territories is extremely complicated. The Kuriles' definition is merely one component of a complex political creation. It was depending on their national interests and posture in the international relations of these periods that the United States, Japan and the Soviet Union dealt with this problem.

THE END OF COLD WAR AND GORBACHEV'S VISIT TO TOKYO: THE THIRD CHANCE FOR *RAPPROCHEMENT*

During the post-war period, Japan had relatively low priority in Soviet foreign policy (Hara 1991: 3–5). No Soviet General Secretary had ever visited Japan, and for ten years – from 1976 to 1986 – no Soviet Foreign Minister visited Japan either. Since Japan was not a military power, and was viewed by the Soviets as only a semi-independent country fully within the US sphere, the Soviets were not interested in any blandishments toward Japan, which seemed unlikely to change Japanese policy or public opinion.

However, since Mikhail Gorbachev became the Soviet political leader in 1985, the Soviet attitude toward Japan changed remarkably under his 'New Thinking Diplomacy'. With the Soviet policy shift toward reducing the burden of military expenditure and aiming at economic recovery, Japan's importance greatly increased for the Soviets. The USSR came to realize the necessity to change its indirect and non-comprehensive policy toward Japan. Soviet attention was directed toward Japan as Gorbachev began to search for a new economic model. Furthermore, Japan's growing military strength, commitment to protect sea-lanes and capability of blocking the straits in the event of war caused the Soviet Union to rethink its attitude towards Japan. The change of direction in Soviet policy became clear with Gorbachev's two famous speeches in the Far East, at Vladivostok (1986) and Krasnoyarsk (1988). In these statements he declared Soviet intent to play a broader and more active role in the Pacific, and acknowledged 'Japan has turned into a power of front-rank importance'.[4] He cited the need for economic development of the Soviet Far East, for which improvement of relations with Japan is a necessary prerequisite. Additionally, at the 1990 Conference on Security and Co-operation in Europe, he made specific mention of Soviet–Japanese relations and emphasized the importance of improved relations between the two nations.[5]

In April 1991 President Mikhail Gorbachev became the first incumbent leader to visit Japan throughout the history of Russia and the USSR. The long-anticipated visit was thus, in itself, a highly significant event. During his visit, Gorbachev addressed the Diet, held more meetings with Prime Minister Toshiki Kaifu than originally scheduled, and, as expected, signed a number of agreements on bilateral exchange and co-operation. The end of the Cold War inevitably, though slowly, began to change the deep-seated antipathy towards the USSR which had long been a hallmark of Japan's foreign policy. Soviet–Japanese relations entered a third period of *rapprochement* – not in a period of temporary *détente*, but against the background of the end of the Cold War.

Gorbachev's visit and the positive personal impact he made on the Japanese people helped improve Moscow's image in Japan. Within the USSR, glasnost permitted various proposals for solving the dispute to be discussed publicly – a marked change from the past, when only the traditionally hardline voice of the Soviet government was heard. The proposals included a transfer to Japan of the Habomai islands and Shikotan, the demilitarization of the islands, the construction of a joint economic zone on and around the islands, their placement under UN trusteeship and so on. Public articulation and discussion of such proposals would have been unimaginable ten years earlier.

Preceding Gorbachev's visit, formal discussions on the territorial question had started in December 1988 when the Soviet Foreign Minister, Edward Shevardnadze, visited Tokyo. Subsequently, official talks between the two sides were held many times, and prospects for a solution to the territorial issue seemed to improve as late as mid-1990. The end of the Cold War, the flexible Soviet attitude toward German reunification, Moscow's normalization of relations with China and its overtures to South Korea – all suggested a new Soviet flexibility and willingness to compromise that augured well for progress on the territorial question.

However, by the time of his visit to Japan Gorbachev's position was already weak, and the Soviet position was not so flexible on the territorial issue as it looked in Soviet international affairs under the earlier Gorbachev regime.

Gorbachev's rival Boris Yeltsin claimed that the disputed islands, which are under the jurisdiction of his Russian Republic, could not be transferred without the agreement of the people of the Republic. Public opinion became a new factor in Soviet and Russian politics. In addition to purely bilateral considerations, political leaders had now also to contend with the fact that the people were adamantly opposed to returning the islands, which they saw as Soviet territory and on which thousands of Soviet citizens live. To further complicate matters, within the Russian Republic, the governor of the state of Sakhalin, Valentin Fyodorov, whose territory includes all the Kurile islands, demanded that *it* be included in the negotiations over the future of the territories. Many Soviets, especially the conservatives, were also concerned that relinquishing the disputed islands could create a dangerous precedent at home – encouraging independence movements throughout the country and thus risking the political disintegration of the USSR (which eventually occurred anyway).

There is little doubt that Gorbachev would have had fewer domestic problems in gaining a settlement with Tokyo had he sought to do so back in 1986 or 1987 – before democracy and glasnost had begun to take hold in the USSR – rather than in 1991 when his political freedom to manœuvre was already sharply constrained. Thus, in terms of Soviet domestic politics, the timing of Gorbachev's visit was extremely bad. The conflict within the USSR on the territorial issue was reflected in the political composition of the delegation Gorbachev brought with him – a mixture of conservatives, reformists and representatives from the republics (ironically the delegation contained no Japan specialist).

There were constraints on the Japanese side as well. Prior to Gorbachev's visit, Prime Minister Kaifu flew to the United States to confer with President Bush. Gorbachev's visit had been a source of both interest and concern in Washington and Japan was, as always, attentive to US concerns. It was very clear that the United States did not want any improvement in the Tokyo/Moscow relationship to affect the smooth running of the US–Japan alliance. Major concessions to Moscow could have strengthened the USSR's diplomatic influence in the region, undermined the US arguments that its allies should spend more on defence and lent credibility to Soviet proposals for improving regional security. The United States saw the latter as intended to undermine its security role in the region and hence rejected them. Aside from this, 'hard-liners' within the Japanese bureaucracy seemed to favour an all-or-nothing approach, complicating Kaifu's position.

On the positive side of Gorbachev's visit, the two sides did agree to work toward a peace treaty based on 'all positive developments' made since the 1956 Soviet–Japanese Joint Declaration. Furthermore, Etorofu, Kunashiri, Shikotan and the Habomai islands were specified in the communiqué for the first time. The future of all the disputed islands was put on the negotiating table; but contrary to the expectations of some observers, the Soviets refused to consider the transfer of Shikotan and the Habomai islands to Japan, which had been promised in 1956. Moscow did, however, agree to permit Japanese citizens to visit all the disputed islands without visas and to reduce the Soviet military presence on the islands. As to the Soviet proposal to have an Asia–Pacific version of the Conference on Security and Co-operation in Europe, the Japanese response was predictably negative. The security situation in the Asia–Pacific region was both different from, and far more complicated than, that in Europe, Tokyo argued.

Many commentators at the summit expressed disappointment at the lack of significant progress on the territorial dispute. Yet most media expectations were almost certainly too high at the outset. The true significance of the summit was that it demonstrated that the territorial issue is no longer an impermeable barrier to improved relations between Tokyo and Moscow. Relations started to improve on other fronts. This overall improvement in the Tokyo/Moscow relationship is the important and positive legacy of Gorbachev's visit.

NEW RUSSO-JAPANESE RELATIONS AND THE TERRITORIAL ISSUE

In August 1991, four months after Gorbachev's visit to Japan, a coup was attempted by Soviet conservatives. It failed in short order, but sealed the fate of Gorbachev's regime, which had been locked in tension between conservatives and reformists. Even after the breakdown of communist rule, Gorbachev strove for the survival of both his regime and the disintegrating Union by forming a loose federation among Soviet republics. It was not to be realized; instead, with the formation of the CIS in 1991, the USSR came to the end of its sixty-nine-year history.

Ironically, Gorbachev became the first and last Soviet leader ever to visit Japan. However, the chance he created for a new era in bilateral relations has not necessarily disappeared along with the Soviet Union. It has been succeeded by Russia, under the leadership of Boris Yeltsin, yet bilateral relations and progress on the territorial question will be considerably affected by the political changes associated with the Soviet disappearance, in both the short and the long term.

In the long term, this historic political change will probably serve as a positive influence on negotiations for a peace treaty and the territorial issue. The Cold War has vanished both in ideology and in political form. This means that the foundation of post-war impediments to bilateral talks has been removed. Since the territorial problem was created largely as a by-product of the Cold War, this political change may well be a real key to resolution of the issue. The independence of the Baltic republics has already created precedents of this kind.

On one hand, the end of the Cold War and disappearance of the USSR has opened new possibilities for a breakthrough. On the other hand, however, the growing complexity inside the Russian Republic may lead both sides to be more prudent about making any concessions. That is, the chaotic situation in the Russian Republic will remain an impediment to progress for the time being.

Political unrest within the Russian government (that is, conflict among various factions within the cabinet and parliament) is already coming to the surface. The economic situation is also not likely to improve quickly, and will remain a major source of social unrest. In a time when there are even rumours of the disintegration of Russia itself, tensions not only among different ethnic groups but between centre and frontier are becoming conspicuous.

The Japanese position

Since the failure of the coup attempt, the focus of Japanese policy toward the former USSR and Russia seemed to be shifting toward co-operation in assisting their democratization and conversion to a market economy. In the light of a series of progressive proposals or ideas regarding the territorial question,[6] the Japanese government appeared to view the problem in the broader context of building a constructive relationship with Russia. To this end, Japan initiated programmes of humanitarian assistance through both public and private channels.

However, at the January 1992 Co-ordinating Conference on Assistance to the New Independent States, where Japan announced its plan to provide about $50 million in food and medical assistance, the Japanese Foreign Minister, Michio Watanabe, reminded the international community of Japanese policy principles for full-scale financial assistance to Russia – 'a shift to a market economy, further progress of democratization and the conduct of diplomacy on the basis of law and justice'.[7] The last principle indicates the return of 'Northern Territories'. He also said 'immediate assistance for the CIS should be limited to humanitarian purposes'. This seems to retreat from the Japanese position since August 1991. Fearing a 'blank period' in the territorial negotiations due to the political chaos in Russia, the Japanese government probably decided to advance its concerns at this time.

Thus, while offering some assistance to Russia, based on the understanding that it is the overall improvement in relations that will make an eventual solution of the territorial issue more likely, Japan is nevertheless keeping to its principle that the solution of the Northern Territories problem is a prerequisite to full economic assistance.

The American role

Like Japan, the United States has not started full-scale financial assistance to Russia, except for humanitarian purposes. When President Bush visited Japan in January 1992, the United States announced that it will 'continue to give maximum support to Japan's efforts to resolve the Northern Territories issue'.[8] In view of its historical involvement and responsibility, its present importance in international political leadership and its global partnership with Japan, it is natural that the United States play some political role in the issue. Since the United States is becoming much closer to Russia in the aftermath of the Cold War, its role may be the positive one of encouraging a normalization of Russo-Japanese relations.

At the time of Gorbachev's visit to Japan, US concerns over the political and strategic situation in the North Pacific led it to oppose Japanese concessions to the Soviet Union. However, circumstances seem to be changing fundamentally and rapidly in the region. On 1 January 1992 Vladivostok was completely opened to foreigners. The opening of this major strategic city in the Far East, where the headquarters of the former Soviet navy is located and from which foreigners had long been banned, verifies that the end of the Cold War had finally reached the region. Furthermore, Yeltsin announced that he intended to stop targeting US cities with nuclear missiles and no longer considered the United States a potential enemy.[9] These remarkable changes in political and security circumstances allow the United States a lot more flexibility in dealing with Russo-Japanese relations.

Russo-Japanese relations have critical meaning to US policy in the Asia–Pacific region. A major US concern is the stability of Russia. Though the United States cannot financially afford full-scale support by itself, it has strong political influence on Japan, which has the financial capability and is now seeking ways to share more of the burden of international contributions for peace. The US political influence is growing on Russia, which is now desperately trying to shift to a market economy and promoting democratization. Thus there is a strong possibility that the United States can be a mediator between Russia and Japan, which are still in an impasse on territorial problems. A breakthrough in the Northern Territories problem will be the key to establishment of a new order in the Asia–Pacific region, and to economic development in the area, both of which will benefit the United States in the long run.[10]

TOWARD THE SOLUTION OF THE TERRITORIAL PROBLEM

Though the activities of the working group for a peace treaty were suspended due to political confusion in Russia, the first negotiations between Japan and Russia started in February 1992 in Moscow. To ensure some concrete progress, active political talks between the two nations are necessary.

When the Northern Territories problem is seen as a vestige of the Cold War between East and West, the necessary condition for its long-term solution seems to be ready. The basis of impediments to constructive talks between Japan and Russia has been removed. The United States has the capability to be a mediator between the two, which will be beneficial to relations between Japan, Russia and the United States in the future. There is great potential for a solution to be found via cooperative improvement of the trilateral relations, as a part of the process of forming a new order in the Asia–Pacific region.

There are many hurdles to overcome in the short run, however. The principal difficulties involve uncertainty about political stability in Russia. Under present conditions, the progress of Russo-Japanese negotiations will largely depend on the Russian side. How Yeltsin overcomes domestic problems will have a major effect on the solution of the Northern Territories issue. As was the case with Gorbachev, the domestic political landscape is not stable enough to allow Yeltsin much flexibility on the territorial issue.

Presently there are approximately 25,000 Russians living in the disputed islands; that far exceeds the number of Japanese on the islands at the end of World War II. The fate of people living on those islands cannot be ignored. According to an opinion survey conducted in Russia in January 1992, 71 per cent of the Russians are against the return of the islands. Furthermore, it will be necessary to keep a close watch on the position of the state of Sakhalin with regard to these islands. Governor Fyodorov has strongly opposed return of the islands to Japan and promoted his own proposals, which include economic co-operation and demilitarization of the region. There are already strong counter-proposals by other political leaders regarding the islands' problems within the same state.

What Japan and Russia can do now may be to prepare public opinion in both countries for the solution of the issue. The fact that constructive and flexible discussion on the territorial problem has been continuing among various opinion leaders is a positive sign for the future. However, both sides have to search for mutually acceptable solutions. For this, it is most important that each side deepen its understanding of the other's position. Russia needs to find some internal consensus on the issue, especially between the centre and the region, in the first place. Meanwhile, Japan could shift the focus of its humanitarian assistance to the Far East and seek proposals which will be beneficial for residents of the islands, as well as promoting overall improvement in its relations with Russia.

NOTES

1   Senkaku Retto is located in the East China Sea about 190 km. north-east of Taiwan, 300 km. south-west of Okinawa and 350 km. east of mainland China. It consists of five uninhabited coral islands and three sunken rocks.

2 Takeshima is located in the Sea of Japan, 30 km. east of South Korea's Ulleung Do and 50 km. north-west of the Japanese Dogo. It consists of two tiny uninhabitable islets and several nearby sunken rocks attached to them.

3 In January 1981 on the basis of resolutions adopted in the previous year, Japan's Cabinet designated 7 February – the date of the signing of the Russo-Japanese Treaty of Commerce, Navigation and Delimitation (Treaty of Shimoda) in 1855 – as 'Northern Territories Day' (Ministry of Foreign Affairs, Japan, 1987: 21).

4 Speech by Mikhail Gorbachev in Vladivostok in 1987 (*Far Eastern Affairs*, 1: 13).

5 'The arch of co-operation we are beginning to develop around the Northern Hemisphere would be inconceivable without developing a new relationship between the Soviet Union and Japan' (Gorbachev, quoted in *Japan Times*, 21 November 1990).

6 For example, G. Popov, the Mayor of Moscow, said the four disputed islands should be returned to Japan (*Japan Times, Asahi Shinbun*, 13 October 1991), and G.F. Kunadze, Deputy Foreign Minister of Russia, stated his opinion that the USSR should fulfil the international agreements of the past and use the 1956 Joint Communiqué as their starting-point for negotiations (*Sankei Shinbun*, 19 October 1991).

7 *Japan Times*, 24 January 1992.

8 The Tokyo Declaration, 'On The US–Japan Global Partnership, Plan of Action (Part I)', *The Daily Yomiuri*, 10 January 1992.

9 *Japan Times*, 27 January 1992.

10 For an interesting argument from a Russian perspective along this line, see Ivanov (1991).

REFERENCES

Asahishinbun Hoppō Ryōdo Shuzai-han (1991) *Hoppō Yontō*, Tokyo: Asahishinbun-sha.

Byrnes, J.F. (1947) *Speaking Frankly*, New York: Harper & Brothers Publishers

Gorbachev, M. (1987) *Far Eastern Affairs*, 1: 13.

Hara, K. (1991) 'Kuriles Quandary: The Soviet–Japanese Territorial Dispute', *Pacific Research*, 4/2.

Hatoyama, I. (1957) *Hatoyama Ichiro Kaikoroku*, Tokyo: Bungeishunju-sha.

Hellmann, D.C. (1969) *Japanese Foreign Policy and Domestic Politics: The Peace Agreement with the Soviet Union*, Berkeley and Los Angeles: University of California Press.

Irie, K. (1962) 'Senryo kanrika no hoppō ryōdo', in Ohira (ed.), *Hoppo ryodo ro Chii*, Nanpōdōhōengo-kai, Japan.

Ivanov, V. (1991) 'The Dynamic Triangle', *Far Eastern Economic Review*, 14 November: 25.

Johnston, D.M. and Valencia, M. (1990) *Pacific Ocean Boundary Problems: Status and Solutions*, Dordrecht/Boston/London: Martinus Nijhoff Publishers.

Kimura, H. (1989) *Hoppo ryodo: kiseki to henkan eno joso,* Tokyo: Jijitsushin-sha.

Kubota, M. (1983) *Kuremurin eno Shisetsu: Hopporyodokosyo 1955–1983,* Tokyo: Bungei shunju.

Matsumoto, S. (1966) *Mosukuwa ni kakeru niji: nisso kokko kaifuku hiroku,* Tokyo: Asahishinbun-sha.

Ministry of Defence, Japan (1991) *Heisei sannendoban Boei Hakusho (Defence of Japan, '91),* July.

Ministry of Foreign Affairs, Japan (1987) *Japan's Northern Territories.*

National Movements Liaison Council for Demanding the Return of the Northern Territories (1987) *Why Is there no Peace Treaty between Japan and the Soviet Union?,* Japan.

10

ITU ABA ISLAND AND THE SPRATLYS CONFLICT

Peter Kien-hong Yu

INTRODUCTION

In the last nineteen years the South China Sea has witnessed many a conflict. On 15 January 1974 the then Republic of Vietnam (ROV) naval vessels fired on fishing boats of the People's Republic of China (PRC) as well as naval ships in the vicinity of Robert Island (*Ganquan Dao* in Chinese) in the Paracel Island group (*Xisha Qundao* in Chinese and *Hoang Sa* in Vietnamese). Two days later Saigon's troops landed on Robert Island and tension escalated. On 19 January the PRC and the ROV fought against each other over the possession of what each claims to be its sacred territory – the entire Paracels – for two days. Beijing dispatched twelve naval ships, four fighter-planes and 500 marines and succeeded in driving out the South Vietnamese from the island group once and for all (*WWP*, 4 October 1988: 9; *NMS*, 1988: 21); but that was not the end of armed conflicts between the two sides over the Paracels.[1]

In another historic encounter, the PRC and the Socialist Republic of Vietnam (SRV) clashed against each other on 14 March 1988 for 28 minutes. The incident took place in Chigua Reef (*Gac Ma* Island in Vietnamese) of the Jiuzhang (*Sinh Ton* in Vietnamese) Island group. The outcome of the clash was, again, in Beijing's favour: the PRC sank one Vietnamese supply-boat and set fire to one landing craft and another supply-boat; only one of its personnel was wounded on Chigua Reef (*RMRB*, 1 April 1988: 1; *WWP*, 1 April 1988; *CN*, 21 April 1988: 2).

As of today, the Republic of China (ROC) on Taiwan is in possession of Itu Aba Island (*Taiping Dao* in Chinese).[2] It has also planted its flags on other islands but the flags were removed by the Vietnamese.[3] Since January 1988 the PRC has landed troops on at least eight islands. The SRV since April 1975 has controlled more than twenty islands and gave up a few (*TKP*, 4 June 1988: 2; *AB*, July 1988: 87).[4] The Philippines since 1971 has occupied at least seven islands. Malaysia is in control of three islands and claims the southern portion of the Spratly archipelago. In late 1990 Brunei released a new map of its coastal jurisdiction and sovereignty. To put it another way, all the above-mentioned countries or political entities are having disputes with each other, and Indonesia may also be involved because it is having problems with the SRV over the continental-shelf issue.

After the March 1988 incident a reporter for the Hong Kong-based *South China Morning Post* said 'the brief encounter . . . has jolted many a nation to face up to the stark realities of the situation' (19 March 1988: 16). Indeed, the situation the ROC is in after that clash is the subject of this study. The first part of the chapter will examine the issue from the military, political and economic perspectives; in the concluding section, the discussion will focus on the applicability of Itu Aba Island's situation to the island of Taiwan itself.

THE MILITARY PERSPECTIVE

Historically, the Spratly Island group belongs to China. From December 1946 to May 1950 and then from 1956 up to now, possession of Itu Aba Island by the Nationalist Chinese has kept the Chinese claim to the entire Spratlys alive. In September 1975 Beijing's leader, Deng Xiaoping, reminded a visiting high-ranking official from the SRV that the Spratlys belong to China (*RMRB*, 25 March 1988: 4). From 1976 the PRC began to patrol the area (*PP*, June 1988: 8). The Chinese claim was further reinforced by the request from the United Nations to the PRC to set up two weather observatories (Nos 74 and 76) in the Spratlys in March 1987 and by the PRC's grand-scale naval cruise from the north-east Cay (*Beizie Dao* in Chinese) in the north to James Shoal (*Zenmu Ansha* in Chinese) in the south from 6 May 1987 to 6 June 1987 for the first time. In January 1990 a group of ROC officials (from the Ministry of the Interior, Ministry of Foreign Affairs and Ministry of National Defence), scholars and newsmen sailed to the Spratlys to erect a monument on Itu Aba Island.

The implication for the change in the situation is that the burden of defending the Spratly archipelago has been shifted from the ROC to the PRC since 1987, when a PRC flag was planted on the island group for the first time. Thus the People's Liberation Army navy has been educating its officers and soldiers in understanding the law of the sea and so on (*JFQB*, 2 July 1988: 1).

184

From the PRC's actions, the security of Itu Aba Island, which is equipped with modern electronic surveillance instruments, seems assured in the foreseeable future, as it has been in the past; but, for the sake of argument, would the 500 ROC marines on the remote military outpost who have been instructed to avoid any conflict with naval vessels from other contestants, including the PRC, and who have been put on full alert since the March 1988 clash, be able to defend themselves from attack? The answer seems negative, as will be explained below (*FEER*, 17 March 1988: 24; 5 May 1988: 26; *CN*, 15 March 1988: 1).[5]

First, the ROC navy faces a logistics problem in conducting a continuous sea-battle from its southern naval port of Gaoxiong, as do the SRV and the PRC from their respective nearest naval bases. Lacking aircraft carriers is the major problem (*SCMP*, 15 March 1988: 1; *WWP*, 12 April 1988: 7).[6] It takes about three days and nights from Gaoxiong to Itu Aba Island by destroyer. A convoy of three ships – usually one destroyer, a tanker, and a supply ship – visits Itu Aba Island to replenish troops' rations and ammunition and to rotate military personnel three times a year (*FEER*, 5 May 1988: 26).[7] At other times, chartered fishing boats would accomplish the mission once or twice a month (*UDN*, 13 April 1988: 1). Given the unusual situation in 1988, two destroyers and a few supply ships were dispatched to the remote outpost in August of the same year (*UDN*, 22 August 1988: 1). In addition, the ROV navy does not have the experience of cruising around the Spratlys in a large formation, even though a reading of an ROC navy chart published in April 1988 suggests that Itu Aba Island and its vicinity, for example, are well surveyed (thus an explosives dumping-ground not far away from the island is revealed).[8] In this connection, it should be noted that the ROC has only twenty-six destroyers, most of which are of the World War II vintage. Its newly acquired conventional submarines (one in December 1987 and the other in July 1988) from the Netherlands can be deployed to the Spratlys. The PRC has sent submarines to the Spratlys including Xia-class nuclear submarines (*WWP*, 10 April 1988: 9; *SCMP*, 7 April 1988: 1).[9] Nevertheless, it is doubtful that the ROC will do so in the foreseeable future, in view of the fact that it has to be able to operate them in the waters surrounding the Taiwan area first so as to cope with any threat from the PRC, including those from the armed fishing vessels. To be sure, only nine ROC naval ships normally patrol the Taiwan waters (*TSWDN*, 20 October 1988: 10). In short, the logistics problem is very unlikely to be resolved, because, first, in the eyes of the ROC's naval planners, defending Taiwan is more important than defending the military outpost[10] and, second, its navy plans by 7 October 1999 to finish building eight Perry-class frigates equipped with harpoon/standard missiles, not warships which have longer cruise range.[11]

Second, defence of Itu Aba Island requires air-support; the ROC lacks that. The island does not have an airstrip as many people have said (*The Nineties*, May 1988: 84; *FEER*, 5 May 1988: 26; Ko 1986: 25). Even if it has, it is of practically no use because the ROC does not possess aircraft carriers. There are, however, a few places on the island that can land helicopters. Thus in mid-1963 the ROC marines received assistance from a US helicopter rescue mission (*TSWDN*, 9 September 1988: 66). The air-support problem is also shared by the PRC but, first, its MiG-19s, H-6 Badger bombers and other military transport can reach the Spratlys, although they are no match for the SRV's MiG-23s and Su-20/22 fighters (*CP*, 17 March 1988: 1; *FEER*, 5 May 1988: 23; *JFQB*, 21 May 1988: 4); second, it has been designing aircraft carriers for many years already;[12] and, third, it had finished constructing a helicopter landing-pad by mid-1988 on Fiery Cross Reef (*Yonshu Jiao* in Chinese and *Chu Thap* or *Chau Vien Reef* in Vietnamese; *SCMP*, 29 February 1988: 9; *TKP*, 31 July 1988: 1). It is claimed that Beijing can construct an airstrip on this 140-square-mile reef if it chooses to (*TKP*, 31 July 1988: 1), although it can use Hainan Province as a staging-base and Woody Island (*Yongxing Dao* in Chinese) of the Paracels as an intermediate base. Vietnam has an airstrip on Spratly Island (*Nanwei Dao* in Chinese and *Truong Sa* in Vietnamese) which is about 500 metres long and which can handle small planes such as Soviet-built CN212 light transport. This airstrip can also be defended by MiG-19s (*CP*, 17 March 1988: 1). According to another report, Hanoi began to convert Bojiao into a naval base with a 450-metre paved airstrip in 1987 (*CN*, 8 August 1988: 1). Manila also maintains an airstrip on Thi Tu Island (*Zhongye Dao* in Chinese and *Pagasa* in Filipino) long enough to land a C-130 cargo plane, and it takes about two-and-a-half hours by flight from the Philippines to this main base (*SCMP*, 23 March 1987: 21). Malaysia now has a small 'hotel' and a helicopter landing-pad on Swallow Reef (*Terumbu Layang-Layang* in Malay and *Danwan Jiao* in Chinese).[13] In short, Itu Aba Island is very vulnerable to aerial attack.

There is a third reason why the ROC troops on Itu Aba Island are vulnerable to attack. While the strategy adopted by Manila in case of 'invasion' to its 'Freedomland' or 'Freedom Island' (*Kalayaan* in Tagalog) is to let the enemy come in as its troops initiate 'a retrograde movement and encircle them later' (*MB*, 24 March 1988: 8), the ROC can not do so. Once there is a retreat, Itu Aba Island will be immediately occupied by other powers. Unless it is the PRC which fills the power vacuum first, the Chinese claim to the entire Spratlys will suffer in one way or another. The scramble for occupation may also spark a large-scale war in the South China Sea.

Finally, unlike the PRC which has operated a satellite capable of identifying combatants at sea since December 1986 and the SRV which was aided by the then Soviet Union after the March 1988 clash, the ROC does not possess a satellite of any kind, be it for surveillance, or signal collections, or radar, or weather or navigation.[14] Of course, we cannot rule out the possibility that Taipei is secretly co-operating with Washington in getting the necessary information. What is sure is that the March 1988 incident was filmed by ROC troops and monitored word by word.[15]

THE POLITICAL PERSPECTIVE

As can be seen, Itu Aba Island is militarily vulnerable. Thus it was rumoured that the outpost fell into the hands of hostile powers in May 1987 (CT, 28 May 1988: 2). It is also vulnerable during the typhoon season (June to October and mid-July to late October) to an invasion (MP, 12 May 1988: 8; WWP, 25 May 1988: 2). But politics, as the art of the possible, can transform something dangerous into something peaceful and tranquil. As such, can Taipei seek help from Washington to deter hostile powers?

The answer seems to be a 'no', unless the PRC with its preponderant power granted its permission overtly or covertly. To be sure, Washington kept a neutral stand during the March 1988 incident (CT, 16 March 1988: 3). This is a long-held US position. On 10 September 1968 the State Department stated that 'the United States Government considers the sovereignty to be undetermined over the islands and reefs which constitute the Paracel and Spratly islands' and in April 1990 a Pentagon report entitled 'A Strategic Framework for the Asia–Pacific Rim' listed the Spratly Island group among other places as an 'unresolved' area.[16] So Taipei cannot rely on Washington politicians or the Taiwan Relations Act (TRA) to deter hostile powers in the Spratly Island group, just as Manila cannot receive help from Washington in the island group (CD, 2 November 1992: 9).

Regarding the former Soviet Union, it is no secret that it sided with the SRV following the demise of the ROV and, therefore, it was inconceivable that it would help the ROC. As early as 3 September

1980, the two communist powers had signed an agreement to explore oil and natural resources in the Spratlys; but it is interesting to note that, when three US navy fliers crash-landed in the Spratlys on 12 July 1988 after their transport plane ran out of fuel *en route* from Singapore to the Philippines, Radio Moscow in its Chinese broadcast on 20 July 1988 used the Chinese version of 'Nansha' Island group when referring to the Spratlys (*CN*, 8 August 1988: 1; *WWP*, 23 August 1988: 2). In any case, the only indirect assurance that Taipei gets from Moscow regarding a peaceful solution to the problem is that Moscow is of the opinion that all parties to the Spratlys dispute should negotiate with each other (*The Nineties*, May 1988: 82). This may well be because in recent years the former Soviet and the Russian Federation leaders have been preoccupied with their own internal economic and racial problems. The ROC has been less concerned about a Vietnamese invasion of Itu Aba Island following the break-up of the former Soviet Union and subsequent weakening of the Hanoi–Moscow ties.

As to Beijing, the situation is quite interesting. Before the March 1988 incident questions such as whether there might be a PRC attack on Itu Aba Island were raised (*SCMP*, 28 March 1988: 9); but after the incident questions such as how the two sides of the Taiwan Strait should co-operate with each other or how the PRC should help the ROC to fortify the military outposts have been raised (*FEER*, 5 May 1988: 26; *WWP*, 6 April 1988: 2; 10 April 1988: 9; *TKP*, 1 April 1988: 2). The idea of making the Spratlys a tourist area has also surfaced (*TKP*, 1 August 1988: 3). A courtesy call on officials of the PRC's National Defence University, People's Liberation Army and State Oceanic Administration in January 1991 gave the strong impression that the two sides of the Taiwan Strait would somehow co-operate with each other on the Spratlys issue. A senior colonel said PRC and ROC marines in the Spratly Island group wave at each other and the PRC side would refer to the ROC side as *guojun* ('national army'). In this connection, since December 1987, Beijing has despatched two battalions of some 1,000 marines to the island group, and Bojiao and Great Discovery Reef (*Daxien Jiao* in Chinese), occupied by Vietnam, have been mentioned as possible targets of a joint attack by both the ROC and the PRC (*CDN*, 1 August 1988: 1; *Guangjiaojin*, November 1988: 32–6). Other targets include Spratly Island, Namyit Island, Sandy Cay, Barque Canada Reef, Thi Tu Island and Swallow Reef (*DN*, 14 September, 1992: 32). Needless to say, such a suggestion helps to strengthen Taipei's position on Itu Aba Island. At a Spratlys conference sponsored by the ROC's Interior Ministry on 15 January 1991, the participants overwhelmingly said the two sides of the Taiwan Strait should co-operate. However, the ROC Government, still confined by its 'Three "No"s' – no contact, no negotiation and no compromise – at the official level, would not flirt with the PRC nor take joint military action unless absolutely necessary. Indeed, according to Liu Ta-tsai, who until December 1990 was the Vice-Admiral of the ROC Navy, now is the time for the People's Liberation Army on its own to drive away foreign troops in the Spratlys, because, after the Cambodian settlement, the SRV would form a coalition with the Philippines, Malaysia and Brunei, and, as for the ROC, it could do nothing because it is militarily weak.

Hanoi has challenged Beijing several times in the Spratlys after the March 1988 incident (*TKP*, 2 August 1988: 3; *JFQB*, 29 July 1988: 2). They again exchanged fire in June 1992. However, it is extremely unlikely that the SRV and the ROC would co-operate with each other politically in deterring a third party from invasion. The ROC has already made it clear that should a request for assistance come from the PRC in the Spratly Island group, it would not hesitate to help (*FEER*, 5 May 1988: 26). The facts that a Taiwan firm concluded a barter-trade agreement with the SRV in August 1988 (*FCJ*, 12 September 1988: 8) – the first such agreement in twelve years – and that flight services between the two countries became regular in 1992 help reduce tensions between the two sides. Given its poor economy, Vietnam does not have the military power to invade islands occupied by other countries in the Spratlys, even though it maintains about thirty naval vessels in the area following the March 1988 clash. What Hanoi needs is peace and, therefore, it has stated that Malaysia, the Philippines and the ROC have agreed to resolve the Spratlys issues through talks (*CN*, 8 August 1988: 1–2). Of course, the list may include Brunei. In this connection, the domino theory as put forward by Hanoi (which means that should the Spratlys come under the control of the PRC, the entire area of south-east Asia would be in jeopardy) is but another indication that Hanoi is playing the role of underdog (*UDN*, 16 May 1988: 1). Thus one Vietnamese observer said that if the PRC does not take action to grab more islands, the SRV would not do it either (*TKP*, 27 August 1988: 2). In a word, Vietnam does not pose much threat to the ROC.

The Philippines is also interested in a peaceful solution to the Spratlys dispute. Agreement has been reached between Manila and Hanoi on this issue in principle (*CT*, 4 April 1988: 2). The two sides will engage in further talks on the issue (*UDN*, 28 September 1988: 1). The Philippines has also received assurance from the PRC that no armed conflict will take place in the island group (*RMRB*, 26 November 1987:

6; *FBIS*, 22 March 1988: 5). Because of this assurance, it is very unlikely that Manila would take Itu Aba Island by force for fear of Beijing's strong reaction: it does not have sufficient power since its economy is still in deep trouble and its government is threatened by communists in the Philippines as well as anti-government forces.

Malaysia also regards portions of the Spratlys as its sovereign territory; but it is in no position to form an alliance with the ROC mainly because it is militarily weak. Its naval vessels have to patrol East and West Malaysian waters. Even if it has submarines, Kuala Lumpur cannot do much to keep what it thinks belongs to it. According to a senior researcher at the Institute of Strategic and International Studies in Kuala Lumpur, the Spratlys will eventually return to China, because other countries do not have the necessary power to keep forever what is at present in their possession. In this connection, Malaysia's foreign minister has also said that the dispute should be settled through talks (*UDN*, 3 July 1988: 1).

As to Brunei and, for that matter, Indonesia and Singapore, the ROC maintains unofficial but very good relations with them. As such, they do not pose much threat to the ROC. The only major worry on the part of Taipei is that in a multilateral negotiation Taipei may be out-voted by them, even if both Taipei and Beijing cast votes.

In summary, the ROC does not have to be worried about attack from other countries or political entities. This would include the PRC because ever since 1984 Beijing has been trying to make Taipei accept the 'one country, two systems' formula, and should it attack Itu Aba Island first, the people on Taiwan would immediately reject the proposal, as there are already about 10 per cent of the total population (21 million) openly clamouring for Taiwan's independence since the lifting of the emergency decree (mistakenly known as martial law) in July 1987.

THE ECONOMIC PERSPECTIVE

There is no question that the Spratlys are rich in tropical resources, fish and other valuable marine products, as well as guano, a good fertilizer. Oil has been found in many areas (*RN*, 23 July 1987: 34; 19 November 1987: 1; *TKP*, 31 May 1988: 3). For example, the submerged James Shoal area is estimated to have a deposit of 12.7–17.7 billion tons of crude oil (*JFQB*, 21 May 1988: 3; 23 August 1988: 2). Because of this potential wealth, many countries have declared economic zones of one type or another: Hanoi declared an Exclusive Economic Zone in May 1977 and incorporated the Nansha and Xisha Island groups in the zone;

Manila declared an Exclusive Economic Zone in June 1979; Taipei declared an Exclusive Economic Zone in September 1979; Jakarta declared an Exclusive Economic Zone in March 1980; Kuala Lumpur declared an Exclusive Economic Zone in April 1980; Brunei declared an Exclusive Economic Zone in July 1993; and Beijing announced its law on its territorial waters and the contiguous zones in February 1992. The PRC still regards major portions of the South China Sea as its internal waters. In the words of one official at the PRC's State Oceanic Administration, because the ROC prior to 1949 also regarded it as such, the PRC is merely going along with the ROC's decision.

To maintain sovereignty, numerous arrests of fishermen by one country or another were frequently made. Vietnam has shot at Hong Kong fishermen (*SCMP*, 7 April 1988: 1; *FBIS*, 21 March 1988: 12). Malaysian naval ships have detained fishing boats from the Philippines and Taiwan (*CT*, 1 June 1988: 12; *AB*, July 1988: 87).[17] Manila's armed fishing boat has killed fishermen from the ROC (*UDN*, 7 July 1988: 13). Since the March 1988 incident, many countries have asked their fishermen to refrain from sailing to the Spratlys. These included the ROC as well as the PRC (*CT*, 11 June 1988: 3; *THWDN*, 4 June 1988: 5; *WWP*, 1 April 1988: 2).[18] Ironically, it is said that more than 270 merchant vessels of one type or another pass through the South China Sea each day since 1980 (*WWP*, 10 April 1988: 9; it may also be pointed out that the Malacca Straits are plied by 1,000–1,500 vessels per day, according to the Malaysia Maritime Agency, *CP*, 16 March 1988: 2), but none of the countries interested in the Spratlys can effectively regulate them.

The only solution to the present economic problem is for all the countries involved to set aside their claims to the Spratlys and work together to develop the island group. Of course, Taipei is not yet ready because it may come into contact with Beijing at the official level unless petroleum companies in mainland China are privately owned, which is quite impossible in the next three years.

In November 1992 the ROC Government Information Office stated the ROC's objective:

In addition to lodging strong protests through appropriate channels, the ROC is willing to seek a peaceful resolution to the dispute over sovereignty through bilateral negotiations with each of the nations concerned.

Before the dispute is resolved, the ROC will consider international cooperation under the premise that the nations concerned want to improve relations with the ROC, and the ROC's sovereignty over the Spratly Islands would not be affected.

First, the ROC would consider cooperating in technical fields where controversy is minimal and where joint interests abound. This category includes navigation safety, pollution of the ocean, natural disasters, sea-borne rescue, oceanographic research and so forth.

Second, the ROC would consider responding positively to 'joint exploitation of the Spratly seas' as proposed by the nations concerned in order to avoid having the area's resources go un-exploited as a result of territorial disputes.

Besides, joint exploitation of resources in the area may help alleviate territorial conflicts and create an environment for peaceful resolution of disputes.

Nonetheless, further studies are necessary to decide whether joint exploitation should be bilateral or multilateral, whether the extent and degree of exploitation should be restricted to the survey of marine resources or extended to cover bio-resources or even joint exploitation of mineral resources.

(*FCJ*, 17 November 1992: 7)

IMPLICATIONS FOR THE ROC

As can be seen, Itu Aba Island is vulnerable to an attack by hostile powers. Can we equate the situation in that military outpost to that of Taiwan? The answer seems negative, because none of the south-east Asian countries poses a military threat to Taiwan. As to the PRC, that is a different matter. So long as the PRC upholds its 'one country, two systems' formula and seeks a peaceful solution and Taiwan remains in the hands of the ruling party, Koumingtang, Taiwan would be safe and sound. On the issue of the Spratlys, blood is thicker than politics. The only linkage between Taiwan and Itu Aba Island is a psychological one. If Itu Aba Island falls into the hands of Chinese communists, it will create an impact upon the people of Taiwan. Of course, if Taiwan goes under, marines on Itu Aba Island will immediately give up their arms; but that will probably not come to pass. Besides, the United States is likely to involve itself to see to it that the Taiwan question, or the China question, is resolved peacefully.

NOTES

1   In 1979 the PRC and the Socialist Republic of Vietnam (SRV) fought again in the vicinity of the Paracels: see *Zhongyue Zaizhan Neimu*: 3–6.

2   According to Chi Tung-hsin, who works for the Taiwan Fisheries Research Institute in Jilong, Taibei County, and who has lived on Itu Aba Island for eight years, the name of this military outpost should have been Nanwei Island (Spratly Island). He has a picture of the stone-marker taken on the island to prove the point (author's research).

3   Author's conversation with Chen Tian-shou, specialist at Gaoxiong Fisheries Administration, Gaoxiong Municipal Government. Mr Chen visited the Itu Aba in August 1988.

4   By September 1993 Hanoi was controlling twenty-five islands in the Spratlys chain. See *Window*, 3 September 1993: 29.

5   The figure of 500 marines was reported (*FEER*, 17 March 1988: 24). According to well-informed sources, there are under 300 troops on the island (author's research).

6   It takes six days and nights from Guangdong Province to Fiery Cross Reef (*JFQB*, 2 August 1988: 2). According to another source, it takes three days and four nights from Wuchan County, Guangdong Province, to Fiery Cross Reef (*TKP*, 2 August 1988: 3). In this connection, according to a military expert: 'Beijing must choose between leaving a small force on the reefs with the risk of not being able to repulse an eventual Vietnamese attack or maintaining a significant garrison with all the problems that raised in terms of supplying food and drinking water for three months' (quoted in *AB*, July 1988: 88). In addition, the living is harsh on the Spratlys. Some twenty PRC marines get one barrel of fresh water every two days (see *JFQB*, 2 August 1988: 2; and *TKP*, 8 June 1988: 2).

7   Manila sends provisions every three months or every forty-five days in case of emergencies (*MB*, 24 March 1988: 8). The PRC replenish troops' rations and other supplies two or three times per month (*TKP*, 8 June 1988: 2).

8   The PRC also completed the surveying of depths distribution of the Spratlys at more than 6,000 spots.

9   According to Chi Tung-hsin, US and Soviet submarines have also been seen in the Spratlys. See *HKT*, 24 March 1988: 3.

10   See Yu 1988a: 30–1. Wang Sheng, former Director-General of the Political Warfare Department, Ministry of Defence of the ROC, is the highest-ranking officer who has ever visited Itu Aba, according to Chi Tung-hsin (author's research).

11   Construction of the first frigate began on 10 January 1990. It takes about forty months to build one frigate.

12   See Yu 1988b: 76–7. See also *RN*, 2 March 1987: 34. Reportedly, the PLA Navy decided to scrap such a project (*UDN*, 11 December 1989: 3).

13   B.A. Hamzah of the Institute for Strategic and International Studies in Kuala Lumpur has confirmed this (author's research). Other news reports have since confirmed that Malaysia is turning Swallow Reef (Terumbu Layang-Layang) into a 'tourist resort'. The Malaysians have maintained a naval presence on the island since 1988 (*FEER*, 20 June 1991: 20; *Straits*

Times, 4 May 1991: 10). On 31 August 1991 Datuk Seri Najib, Malaysia's Minister of Defence, confirmed that Malaysia was developing a runway for light aircraft and installing military equipment on the island to detect air attacks (*Voice of Malaysia,* Kuala Lumpur, radio report on 31 August 1991).

14 *Diyishouziliao,* 38, September–October 1988: 116. See also *JFQB,* 24 April 1988: 4; *UDN,* 17 March 1988: 1.

15 Conversation with Jason C. Hu, Associate Professor at Sun Yat-sen Institute, National Sun Yat-sen University, Gaoxiong, Taiwan (author's research).

16 However, the US Defense Department later retracted that statement (*FCJ,* 26 April 1990: 1–4).

17 The author has testified before the Malaysian court in November 1988 that the Spratlys belong to China and, therefore, detainment of four fishing boats from Taiwan in the Spratlys waters by Malaysia in August 1988 was illegal: see Yu 1990 and 1990–1.

18 Hong Kong fishermen were also warned not to catch fish in the Spratlys waters but some still do (see *WWP,* 4 September 1988: 3).

REFERENCES

Newspapers and periodicals

Asian Bulletin, Taipei (AB).
Central News Agency, Reference News, Taipei (RN).
China Daily News, New York (CDN).
China Information, Holland (CI).
China News, Taipei (CN).
China Post, Taipei (CP).
China Times, Taipei (CT).
Commons Daily, Kaohsiung, Taiwan (CD).
Defense News, USA (DN).
Divishouziliao, Taipei.
Far Eastern Economic Review, Hong Kong (FEER).
Free China Journal, Taipei (FCJ).
Foreign Broadcasting Information Service, USA (FBIS).
Guangjiaojin ('Wide Angle'), Hong Kong.
Hong Kong Times, Hong Kong (HKT).
Jiefangjunbao, Beijing (JFQB).
Manila Bulletin, Manila (MB).
Ming Pao, Hong Kong (MP).
Naval and Merchant Ships, Beijing (NMS).
Pacific Review, UK (PR).
PLA Pictorial, Beijing (PP).
Renminribao, Beijing (RMRB).
South China Morning Post, Hong Kong (SCMP).
Straits Times, Singapore.

Ta Kung Pao, Hong Kong (TKP).
Taiwan Shin Wen Daily News, Gaoxing (TSWDN).
The Nineties, Hong Kong.
United Daily News, Taipei (UDN).
Wen Wei Po, Hong Kong (WWP).
Window, Hong Kong.

Secondary sources

Ko, T.H. (1986) 'China's Sea Frontiers', in T.H. Ko and P.M. Chen (eds), *Sea Lane Security Studies: Some Vital Issues*, Taipei: Asia and World Institute.
Samuels, M.S. (1982) *Contest for the South China Sea*, New York: Methuen.
Yu, K.H. Peter (1988a) *A Study of the Pratas, Macclesfield Bank, Paracels and Spratlys in the South China Sea*, Taipei: Tzeng Brothers Publications.
—— (1988b) 'The PLA Navy', in R.H. Yang (ed.), *SCPS Yearbook on PLA Affairs: 1987*, Gaoxiong, Taiwan: Sun Yat-sen Centre for Policy Studies.
—— (1990) 'Protecting the Spratlys', *Pacific Review*, 3/1.
—— (1990–1) 'A Malaysian Trial of Taiwanese Fishermen Caught in the Spratlys: A Case Study', *China Information*, 5/3.
—— (1993) 'A Grand Strategy for Solving the Spratlys Dispute from a Chinese Perspective', in R.H. Yang (ed.), *China's Military: The PLA in 1992/1993*, Taipei: Chinesre Council of Advanced Policy Studies, distributed in the USA by Westview Press.
Zhongyue Zaizhan Neimu (Inside Story on the Renewed Sino-Vietnamese War) (in Chinese) (no date) Hong Kong: no publisher given.

11

COMPARATIVE OIL AND GAS JOINT DEVELOPMENT REGIMES

Francis M. Auburn, Vivian Forbes and John Scott

INTRODUCTION

The settlement of boundary disputes involving resources has traditionally centred on the demarcation of specific lines dividing the disputed resource area between the states involved.[1] However, modern practice has developed a number of possible alternatives. They range from the limited case of unitization of transboundary deposits[2] to an agreed draft covering the resources of an entire continent.[3] There has been disagreement as to whether customary international law specifically requires positive co-operation for joint development (Miyoshi 1988). The lack of certainty on custom and the growing number of agreements suggest that the focus is best placed on negotiated treaties as indications of the various possible solutions available in the absence of conventional demarcation.

This chapter will focus on existing joint-development texts, with particular attention to the Thailand/Malaysia Memorandum of Understanding 1979,[4] the Japan/South Korea Agreement 1974,[5] the Convention on the Regulation of Antarctic Mineral Resources Activities (CRAMRA) and the Indonesia/Australia Timor Gap Treaty 1989.[6] The purpose is to compare specific issues covered by the agreements and, in particular, to examine the degree to which they contribute precedents for legal-conflict resolution.

THAILAND/MALAYSIA

For the area under discussion see Figure 11.1.

The Thailand/Malaysia Memorandum of Understanding is brief. The area dealt with is only 2,700 square miles and it is not expected to

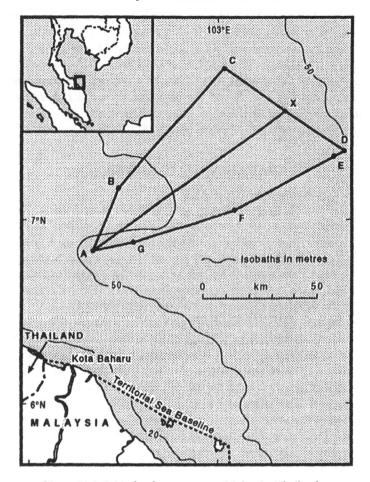

Figure 11.1 Joint development area: Malaysia–Thailand

hold substantial oil deposits (BIICL 1989: 133). The area in dispute is defined by straight lines connecting co-ordinates (Article 1). Both countries agreed to continue to resolve the delimitation problem by negotiation or other peaceful means (Article 2). A Joint Authority was established to explore and exploit the non-living natural resources of the sea-bed and subsoil for fifty years. Existing concessions were not affected (Article 3(2)).

Where oil or natural-gas deposits extend beyond the limits of the joint development zone, the Joint Authority and the party concerned are to seek to reach agreement (Article 3(b)). National rights over fishing,

197

navigation, hydrographic and oceanographic surveys and prevention and control of marine pollution should extend to the joint development area. Security arrangements were to be combined and co-ordinated (Article 4). Criminal jurisdiction was divided between the parties along a specified line; but this line should not be construed as a continental-shelf boundary-line and should not prejudice sovereign rights (Article 5). If no satisfactory delimitation is agreed upon after fifty years, the existing arrangement will continue (Article 6(2)).

JAPAN/SOUTH KOREA

For the area involved see Figure 11.2.

Whilst the Thailand/Malaysia agreement is a bare framework, the Japan/South Korea one is far more detailed and complex. The Joint Development Zone provides for division into subzones, each of which are to be explored and exploited by concessionaires of both parties (Article III). Concessionaires enter into an operating agreement covering (*inter alia*) resource sharing, operator designation, sole-risk operations, adjustment of fisheries interest and dispute settlement (Article V).

Exploration rights last for eight years and exploitation rights for fifty years. Five-year extensions may be given as many times as necessary (Article X(3)). The agreement is to remain in force for fifty years and thereafter with three years' notice of termination (Article XXXI). Each party applies its own laws to the natural resources extracted in the Joint Development Zone (Articles XVI and XIX). Prevention and removal of pollution of the sea resulting from activity relating to exploration and exploitation of natural resources in the Joint Development Zone are to be decided by agreement between the parties (Article XX). Damage resulting from such activities sustained by nationals or residents of either party may give rise to court action (Article XXI). Transboundary resources are to be dealt with by consultation between concessionaires (Article XXXIII).

A Joint Commission was established for consultation on implementation of the agreement. The Commission's decisions require consensus of both countries. It meets at least once a year (Article XXIV). Its functions include the recommendation to the parties of measures to solve disputes which the concessionaires cannot settle themselves (Article XXV(1)(c)). Disputes between the parties which cannot be settled by diplomatic means are to be referred to an arbitration board (Article XXVI(2)). Navigation, fisheries and other legitimate activities must not be unduly affected by resource activity in the Joint Develop-

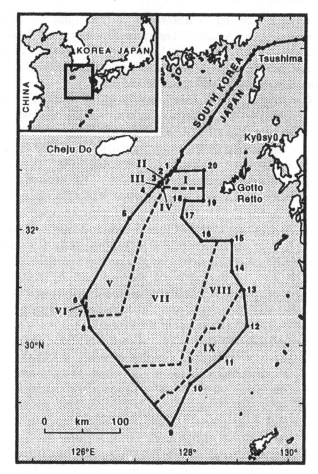

Figure 11.2 Joint development zone: Japan–South Korea

ment Zone (Article XXVII). The agreement does not prejudice the position of either party on continental-shelf delimitation (Article XXVIII).

CONVENTION ON THE REGULATION OF ANTARCTIC MINERAL RESOURCES ACTIVITIES (CRAMRA)

The Convention on Antarctica will be used as a model, rather than for its application in practice, because it has been superseded by a ban on mineral-resource activities in Antarctica in 1992 (also see Article 6, 'Draft Protocol to the Antarctic Treaty on Environmental Protection', XI ATSCM/2/30 (29 April 1991)). However, CRAMRA remains of considerable significance due to the initial agreement of the leading treaty consultative parties and its scope and detail.

The Convention would have applied to all Antarctic mineral-resource activities on the Antarctic continent and islands and the sea-bed and subsoil south of 60°S (Article 5(2)). Nothing in CRAMRA was to prejudice the varying positions of the parties regarding sovereignty (Article 9). An elaborate administrative structure would have provided for no fewer than five different institutions. The Antarctic Mineral Resources Commission would have consisted of all existing consultative parties and other parties actively engaged in resource activities (Article 18). The Commission's functions would include adopting measures to protect the Antarctic environment and promoting safety. It would make budgetary decisions and determine surplus revenues (Article 21(1)).

For each area to be exploited there would be a Regulatory Committee of ten members. Membership would be balanced to include claimants, the two superpowers, non-claimants and developing countries (Article 29). Substantive decisions of the Committees would be made by a two-thirds majority (Article 32(3)). Crucial decisions, including the effective approval of exploration and development, would need a two-thirds majority, including a simple majority both of the four claimants and of the six other members. The Committees' functions would include the approval of management schemes and of permits for exploration and development (Article 31). The management scheme would be central to CRAMRA and would include the duration of permits, performance requirements, liability and applicable law (Article 47). These and other vital issues would be decided in detail by the Regulatory Committees. An Advisory Committee would provide advice on scientific, technical and environmental matters; but that Committee's opinions would not bind the Commission and Regulatory Committees (see, for example, Article 31(2)).

The Convention provides a detailed list of requirements protecting the environment. For instance, no Antarctic mineral-resource activity could take place unless it was judged that no significant changes in the distribution, abundance or productivity of populations of species of fauna or flora would result (Article 4). Environmental-impact assessment would be specifically required for major decisions (Article 26(4)). A special meeting of all parties would advise the Commission whether opening up an area would be consistent with the Convention (Article 40(3)).

TIMOR GAP

The area in question is shown in Figure 11.3.

The Indonesia/Australia Treaty on the Timor Gap 1989 creates a zone of co-operation in the disputed sea-bed boundary between East Timor and Australia covering 60,000 square kilometres (Auburn and Forbes 1991; Forbes and Auburn 1991). Within that zone there are three areas. In Area B (closest to Australia) Australia will pay 10 per

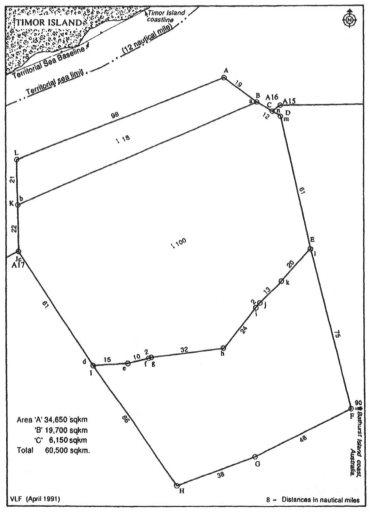

Figure 11.3 The Timor Gap Treaty zone of co-operation

cent of the gross resource rent tax collected from petroleum production to Indonesia (Articles 2(2)(b) and 4(1)(b)). Indonesia makes analogous payments of 10 per cent of contractors' income tax collected in Area C (the area closest to Indonesia) (Articles 2(2)(c) and 4(1)(c)).[7] The largest and central portion, Area A, is subject to a detailed joint-development regime (Figures 11.4 and 11.5).

A Ministerial Council, operating by consensus, has overall responsibility for Area A. It may give directions to the Joint Authority, modify the Petroleum Mining Code and Model Production Sharing Contract, approve production-sharing contracts and perform a wide variety of further supervisory functions (Article 6). The Joint Authority, having juridical personality, is responsible to the Ministerial Council. It acts by consensus (Article 7(4)). It functions subject to the supervision of the Ministerial Council and manages petroleum exploration and exploitation activities (Article 8).

Specific provision is made for co-operation on certain matters in Area A, for instance search-and-rescue, air-traffic services and protection of the marine environment (Articles 14, 15 and 17). The parties are to seek agreement on petroleum accumulations extending across the boundary of Zone A (Article 20). Criminal jurisdiction is generally based on nationality or permanent residence and flag jurisdiction for vessels. Civil actions are generally to be brought in the contracting state whose national or permanent resident has suffered damage (Articles 21(1) and (3)). Dispute settlement between the contracting states is by negotiation (Article 28). The treaty is to be in force for forty years and can be renewed for successive terms of twenty years if no permanent continental-shelf delimitation has been reached (Article 30(1)). A Petroleum Mining Code for Area A deals with bidding, rate of production, work practices, inspection, fees and other specific issues (Article 33(2)). Annex C contains a Model Production Sharing Contract between the Joint Authority and a contractor. There is also a specific Double Tax Agreement (Annex D).

GENERAL COMPARISON

An overall comparison of the four agreements surveyed indicates the close relationship between functional demands and the contents of the agreement in question. The Thailand/Malaysia Memorandum of Understanding dealing with a relatively small area with little likelihood of petroleum is very brief and covers such matters as the functions of the Joint Authority in a very generalized format. The Japan/South Korea

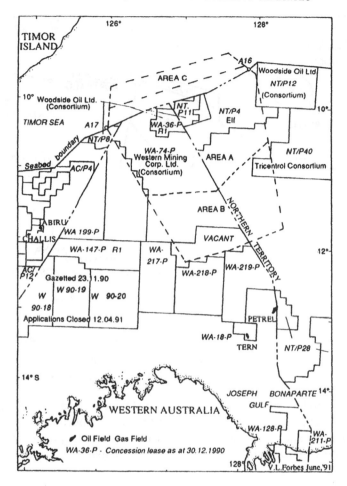

Figure 11.4 Petroleum leases in the vicinity of the zone of co-operation, Timor Sea
Source: Map no. M175, WA Dept of Mines, December 1990.

Agreement goes into greater detail. For instance, its complex system of subzones (Article III) invites difficulties. The Timor Gap Treaty's Zone A contains the Kelp Structure which has been suggested to hold substantial petroleum deposits (Australian Legal Group 1990: 1). This explains the prolonged negotiations and complexity of the treaty. The Antarctic Minerals Convention raised issues concerning sovereignty over a whole continent, the integrity of the Antarctic System and the assumption that the only feasible resource (offshore oil and gas) was

203

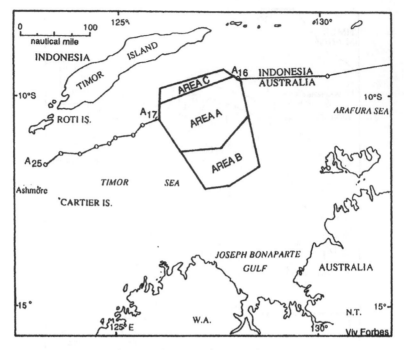

Figure 11.5 Tectonic elements of the Northern Bonaparte Basin
Source: Botten and Wulff (1990), the APEA Journal, pp. 71 (modified).

unlikely to be exploited in the short term (Office of Technology Agreement 1989). These factors accounted for the complex administrative framework and the detailed environmental protection principles. One can now turn to discuss some of the specific comparative issues.

Management

Clearly the structure of administration is crucial to the practical operation of a joint-development resource regime. In the Japan/South Korea Agreement the Commission's role is restricted to carrying out co-operative efforts between the two states. Licensing and regulation remain in the hands of each state. In practice, it has been suggested that such a management system is not entirely satisfactory (Reid 1986/7). Another approach is to have a joint ministerial body deciding policy and a joint corporation exploiting the resources, as in the Aden Agreement between the former separate states of the Yemen Arab Republic and the People's Democratic Republic of Yemen (Onorato 1990). Since this corporation decided to license foreign corporations to exploit (Onorato 1990: 661) the concept became similar to a two-tier proposal.

The Convention on Antarctica provides the most elaborate administrative structure. The Commission is less powerful than the Regulatory Committees (Auburn 1990: 262). The political accommodation between the parties was, in large part, dealt with by the complex membership and voting procedures of the Regulatory Committees (Jacobsen 1989). The Advisory Committee was specifically intended to advise on environmental issues and was given detailed principles to adhere to. The Regulatory Committee would have wide powers, given as headings to be developed in detail for each application. In particular, there would be a separate management scheme for each block (Article 47) with the possibility of differing legal systems and operating requirements for different blocks. Whilst the complicated administrative structure of CRAMRA can be understood against the background of the conflicting interests of the states involved, it would not have been easy to implement in practice. The major reason for the ultimate rejection of CRAMRA was stated to be the weakness of environmental protection (Blay and Tsamenyi 1990). It is therefore not surprising that the 'Protocol to the Antarctic Treaty on Environmental Protection' 1991 added further administrative layers to the Antarctic System, without solving the issues of co-ordination.

The two-tier management structure of the Timor Gap Treaty has been described as unusual (Onorato and Valencia 1990). The close co-operation between the two countries required by the consensus decision-making at each level is both a high level of integration and a problematical procedure. In case of disagreements the entire treaty would be at risk. However, the decision to manage jointly Zone A, without any resolution of sovereign rights, explains the two tiers and the close co-operation at two levels. It remains to be seen whether this will prove satisfactory in practice.

Resources

The Thailand/Malaysia Memorandum of Agreement applies to the exploration and exploitation of the non-living natural resources of the sea-bed and subsoil (Article 3(1)). National rights over fishing, navigation, hydrographic and oceanographic surveys, the prevention and control of marine pollution and other similar matters (including enforcement) extend to the Joint Development Area (Article 4(1)).

The Japan/South Korea Agreement covers natural resources, defined

205

as petroleum, natural gas and other underground minerals produced in association with such resources (Articles I(1) and V(1)). However, operating agreements cover the adjustment of fisheries interests (Article V(1)(e)), apparently in the sense that oil and gas activities will not interfere with fishing (BIICL 1989: 121–2).

The Convention on Antarctica would apply to Antarctic mineral resources (Article 3), defined as non-living natural non-renewable resources, including fossil fuels, metallic and non-metallic minerals (Article I(6)); but CRAMRA required decisions to take into account the need to respect established uses of Antarctica, including the conservation of Antarctic marine living resources (Article 15(1)(c)).

Prima facie, the Timor Gap Treaty deals only with the exploration and exploitation of petroleum and natural-gas resources (Article 2(1)). It does not extend to other minerals (Moloney 1990). In certain instances the further provisions of the treaty are specifically limited to actions done 'for the purposes of this Treaty' (see Article 12 on surveillance activities). In other cases, such as air-traffic services cooperation, the coverage is not limited to the purposes of petroleum (Article 15).

It may be suggested that a co-operative joint-development resource zone may well, in practice, be difficult to limit to the purposes of exploration and exploitation of the natural resources specified. The clearest case arises in treaties including resources on land, such as CRAMRA; but even continental-shelf joint development regimes may well have to expand to cover non-resource activities. The Japan/South Korea fishing provisions are an interesting example, but are not unique. For example, environmental impacts in the water column would generally be covered by such agreements, either directly or by inference.

Environment

The Thailand/Malaysian Memorandum does not specifically deal with environmental issues. The division of the Joint Development Area into two zones of national criminal jurisdiction (Article 5) and the specific reservation of fishing issues and 'other similar matters' to each state (Article 4) suggest that environmental protection of the water column could be seen as reserved to each state.

The Japan/South Korea Agreement provides that the parties shall agree to measures to prevent and remove pollution resulting from exploration and exploitation of natural resources in the Joint Development Area. Damage from natural-resource activities sustained by nationals or residents of either party can give rise to court action in either country (Articles XX and XXI). This provision does not, however, lay down the substantive law and does not protect nationals or residents of other states.

Despite the argument that CRAMRA was rejected because of the weakness of its environmental-protection provisions, the Convention would have provided an elaborate cascade of principles. These would operate at each stage of decision-making. The Convention would have made specific provision for environmental protection. Examples are the establishment of protected areas (Article 13(2)), detailed requirements for environmental assessment (Article 26(4)), safe-operations technology and monitoring capacity (Article 4(4)). It was clear that the effectiveness of CRAMRA's environmental provisions was open to criticism,[8] but, when compared with other treaties for joint development regimes, CRAMRA could be seen as a landmark in environmental protection.

The Timor Gap Treaty provides for co-operation to prevent and minimize pollution of the marine environment generally and that arising from petroleum activities in Area A (Article 18(1)). The Joint Authority's management functions include requesting assistance with pollution-prevention measures, equipment and procedure from the appropriate Australian or Indonesian authorities or other bodies or persons (Article 8(m)).

The Joint Authority was to issue regulations for the carrying-out of operations in an environmentally sound manner (Article 37(k) of the Petroleum Mining Code). Regulations were to be made for environmental-impact assessment (Article 37(1)). The Petroleum Mining Code also requires contract operators to work in accordance with 'good oilfield practice' (Article 24), defined as all those things that are generally accepted as good and safe in the carrying-on of petroleum operations (Article 1(f) in the treaty). Critics question who accepts such practices (contracting states? states generally? industry?) and what 'good' means (Moloney 1990: 174).

Where petroleum escapes, such action is to be taken as is necessary to minimize pollution of the sea (Clause 224, Regulations issued by the Ministerial Council (1991)). Every reasonable precaution should be taken to avoid pollution of the environment (Clause 616). Contractors are liable for damage or expenses incurred as a result of pollution of the marine environment arising out of petroleum operations in accordance with the contractual arrangements with the Joint Authority and the law of the state in which a claim is brought (Article 19 of the treaty). The

Model Production Sharing Contract requires the development of an environmental management plan to be approved by the Joint Authority, prevention of pollution of the marine environment and payment for the cost associated with clean-up of any pollution (Article 5.2(d)). Contractors must insure on a strict liability basis for liability to the satisfaction of the Joint Authority, including expenses associated with the prevention and clean-up of the escape of petroleum (Article 25(1) of the Petroleum Mining Code). The effectiveness of insurance depends on the requirements of the Joint Authority.

The Timor Gap regime attempts to provide detailed rules to prevent, limit and ensure liability for environmental protection. In this it is far more detailed than the Thailand/Malaysia and Japan/South Korea Agreements. However, CRAMRA provides general objectives and principles for environmental protection and detailed requirements for impact assessment which are not yet to be found in the Timor Gap regime.

Existing concessions

The Thailand/Malaysia Memorandum specifically preserves the validity of pre-existing licences (Article 3(2)), but there have been difficulties in its implementation (BIICL 1989: 138–9). The Japan/South Korea Agreement contains no provision for prior concessions, thus indicating that they will cease to be valid. As there were no licences when CRAMRA was drafted, the issue did not arise.

The Timor Gap Treaty adopts the clean-slate approach of not giving recognition to pre-existing rights (Livesley 1990). Before the negotiations Australian authorities had issued licences in Area A (Howarth 1989). This raised the possibility of compensation claims against the Australian Commonwealth Government under section 51(31) of the constitution. The Australian government publicly rejected the possibility that such claims could succeed.

In unpublished side-letters to the treaty Australia and Indonesia allegedly agreed that the Ministerial Council would give favourable consideration to applications by the existing Australian permit holders (Poll 1990: 57). Since the possible losses could amount to hundreds of millions of dollars to the companies concerned (Poll 1990: 76) they have a strong incentive to sue the Commonwealth Government for compensation should a satisfactory solution not be reached. The side-letters offer a solution enabling the Ministerial Council to give preference to existing permit-holders. One of the Australian former permit-holders, the Western Mining Corporation, commenced proceedings against the Commonwealth Government in June 1991 (Pheasant 1991).

However, Indonesia insisted on the clean-slate approach, to reflect its rejection of Australia's sovereign rights in the area (Smart 1990: 387). Preference to existing Australian permit holders, especially in the highly prospective Kelp High, contradicts the clean-slate approach. Relegation of the agreement has the same status, whether it is published or not and whatever its name. It remains to be seen whether and to what extent the permit holders will be compensated by receiving priority in areas equivalent to their original holdings.

The problem of existing permit holders is highlighted by the Timor Gap Treaty. If a total joint development regime is to be adopted, without sovereignty complications, then the clean-slate approach is a logical outcome. On the other hand, no government will be ready to face large compensation claims. The apparent compromise in the Timor Gap Treaty may have helped solve the Australian Government's problems with licences, but this would be at the expense of compromising the sovereignty issue.

Interim regime

Although the agreements are framed as interim regimes, one may ask to what extent they are likely to continue in the long term. The Thailand/Malaysia Memorandum is to last for fifty years and continues after that period unless a satisfactory solution is agreed upon (Article 6(2)). There is specific agreement to continue attempts to delimit the boundary (Article 2).

The Japan/South Korea Agreement remains in force for fifty years and continues in force thereafter, with a proviso of a three years' notice of termination (Articles XXXI(2) and (3)). The Agreement does not contain an undertaking to negotiate to solve the problem.

The Convention on Antarctica differs from the other treaties in that it was intended to be a permanent regime. Parties' legal rights to claim sovereignty, make claims or refuse to recognize claims would have been preserved (Article 9). The Convention would have been part of the Antarctic System which is distinguished by the specific intention of the Antarctic treaty consultative parties not to set up a legal regime for Antarctica (Auburn 1991). The practical outcome in Antarctica has been an open-ended regime of joint control by the consultative parties to the Antarctic treaty. There is no provision in any of the related agreements for a final settlement of sovereignty issues, nor are negotiations on this issue likely in the future.

Nothing in the Timor Gap Treaty and no activities whilst it is in force should be interpreted as prejudicing the position of either contracting state on a permanent continental-shelf determination. The contracting states were to continue their efforts to reach agreement on a permanent continental-shelf delimitation (Articles 2(2) and (4)). The treaty is to be in force for forty years, with a provision for successive twenty-year renewal if agreement had not been reached on a permanent boundary (Article 33(2)).

Discussions on the basic difference between the two parties[9] reached an impasse, forcing agreement in principle on a joint development zone in 1985 (see Wilheim 1989 and Bergin 1990: 384). In view of the inability of the countries to settle the fundamental issue during negotiations over several years, it is unlikely that the situation will change in the future (Agoes 1991).[10] Furthermore, the exploitation of oil and gas under the treaty will provide a diplomatic solution to a situation which neither side has any present interest in disturbing.

It would appear that interim joint-development resource regimes have a tendency to become permanent, especially where the underlying differences between the parties are fundamental and have proved unsolvable in the past.

CONCLUSIONS

The agreements under review provide a wide range of possible solutions to joint development of petroleum resources. The Thailand/Malaysia Memorandum is exceedingly brief. It does not provide for such basic issues as the application of industrial laws. The Japan/South Korea Agreement is more comprehensive, but has encountered difficulties with its provision for alternating operators and national licensing and tax authorities.

The Convention on Antarctica, covering the largest area by far and open to numerous parties, would have been an innovative experiment in joint mineral-development regimes, despite its recognized failings. The fact that twenty consultative parties with widely differing political systems were able to reach agreement on a complex and long-range resource settlement involving sovereignty issues for an entire continent was itself remarkable.

The Timor Gap Treaty was arrived at between two neighbouring countries with very different political systems. It is the most comprehensive joint-development act and gas agreement to date and has been enhanced by the very detailed regulations issued in February 1991. The crucial test of the treaty will be the implementation of its licensing and co-operation provisions in practice.

NOTES

1 The authors have pleasure in expressing their thanks to the Indian Ocean Centre for Peace Studies, the Centre for Commercial and Resources Law and the International Boundaries Research Unit for their support.
2 As in the Frigg Field Agreement between the United Kingdom and Norway (1977), in the *United Kingdom Treaty Series* (UKTS) 113, discussed by Woodliffe 1977.
3 'Convention on the Regulation of Antarctic Mineral Resource Activities' (1988), 27, ILM 868 examined in Auburn 1990: 259.
4 Kingdom of Thailand 1979.
5 'Agreement' 1974.
6 Australian Government 1989.
7 Unless otherwise stated, discussion of the Timor Gap Treaty in this chapter refers to Area A.
8 For example, that the infrastructure would harm the environment, scientific research would be jeopardized and vague value judgements would be needed (see Stephen 1990: 725).
9 Australia relied on geography (the axis of the Timor Trough) and Indonesia on the median line.
10 Negotiations commenced in 1979 (see Agoes 1991).

REFERENCES

Agoes, L.A. (1991) 'Rules of Joint Development: The Timor Gap Treaty', ICLOS/NILOS Seminar, Bandung, Indonesia.

'Agreement between Japan and the Republic of Korea concerning Joint Development of the Southern Part of the Continental Shelf Adjacent to the Countries' (1974): see *Energy* (1981) 6.

Auburn, F.M. (1990) 'Convention on the Regulation of Antarctic Mineral Resource Activities', in J.F. Splettstoesser and G.A.M. Dreschhoff (eds), *Mineral Resources Potential of Antarctica*, Washington, DC: American Geophysical Union.

―――― (1991) 'Conservation and the Antarctic Minerals Regime', *Ocean Yearbook*, 9.

Auburn, F.M. and Forbes, V.L. (1991) 'The Timor Gap Treaty and the Law of the Sea Convention', SEAPOL Workshop, Chiang Mai, Thailand.

Australian Government (1989) *Treaty between Australia and the Republic of Indonesia on the Zone of Co-operation in an Area between the Indonesian Province of East Timor and Northern Australia*, Canberra: Australian Government Printers.

Australian Legal Group (1990) 'Timor Gap and Search', *Headlines*, 3.

Bergin, A. (1990) 'The Australian–Indonesian Timor Gap Maritime Boundary Agreement', *International Journal of Estuarine and Coastal Law*, 5/4.

FRANCIS M. AUBURN *ET AL.*

Blay, S.K.N. and Tsamenyi, B.M. (1990) 'Australia and the Convention for the Regulation of Antarctic Mineral Resources Activities (CRAMRA)', *Polar Record*, 26/158.
British Institute of International and Comparative Law (BIICL) (1989), *Joint Development of Offshore Oil and Gas*, BIICL.
Forbes, V.L. and Auburn, F.M. (1991) 'The Timor Gap Zone of Co-Operation', *Boundary Briefing*, 9.
Howarth, I. (1989) 'Oil Deal Calms Timor Waters', *Financial Review*, 17.
Jacobsen, M.P. (1989) 'Convention on the Regulation of the Antarctic Mineral Resources Activities', *HILJ*, 30.
Kingdom of Thailand (1979) 'Memorandum of Understanding between the Kingdom of Thailand and Malaysia on the Establishment of a Joint Authority for the Exploitation of the Resources of the Sea-Bed in a Defined Area of the Continental Shelf of the Two Countries in the Gulf of Thailand', see *Energy* (1981) 6.
Livesley, K.P. (1990) 'The Timor Gap Treaty', International Bar Association Conference, on 'Energy Law 2000', Darwin.
Ministerial Council exercising the Rights of Australia and the Republic of Indonesia in relation to the Exploration for and Exploitation of Petroleum Resources in Area A of the Zone of Co-operation (1991) Regulations Issued under Article 37 of the petroleum Mining Code, Devpasar.
Miyoshi, M. (1988) 'The Basic Concept of Joint Development of Hydrocarbon Resources on the Continental Shelf', *International Journal of Estuarine and Coastal Law*, 3/1.
Moloney, G.J. (1990) 'Australian–Indonesian Timor Gap Zone of Cooperation Treaty: A New Offshore Petroleum Regime', *Journal of Energy and Natural Resources Law*, 8/2.
Office of Technology Agreement (1989) *Polar Prospects*, 114/1.
Onorato, W.T. (1990) 'Joint Development in the International Petroleum Sector: The Yemeni Variant', *ICLQ*, 39.
Onorato, W.T. and Valencia, M.J. (1990) 'International Cooperation for Petroleum Development: The Timor Gap Treaty', *Foreign Investment Law Journal*, 5/1.
Pheasant, B. (1991) 'WMC Seeking Compo over Timor Sea Rights', *Financial Review*, 14 June.
Poll, J.J.K. (1990) 'Timor Gap Seminar', *APEA Journal*, II.
'Protocol to the Atlantic treaty on Environmental Protection 1991' (1991) 30 ILM (International Legal Materials) 1461.
Reid, P.C. (1986/7) 'Joint Development Zone between Countries', *OCTLR*, 9.
Smart, A. (1990) 'Timor Gap Zone of Co-operation', *APEA Journal*, II.
Splettstoesser, J.F. and Dreschhoff, G.A.M. (eds) (1990) *Mineral Resources Potential of Antarctica*, Washington, DC: American Geophysical Union.
Stephen, N. (1990) 'The Environmental Protection of the Antarctic', *The Monthly Record*, October.
Wilheim, E. (1989) 'Australia–Indonesia Sea-Bed Boundary Negotiations: Proposals for a Joint Development Zone in the "Timor Gap"', *Natural Resources Journal*, 29.
Woodliffe, J.C. (1977) 'International Utilisation of an Offshore Gas Field', *ICLQ*, 26.

INDEX

Printed and bound by CPI Group (UK) Ltd, Croydon, CR0 4YY
01/11/2024
01782616-0007